Nanoscience and Nanotechnology in Engineering

NANOSCIENCE AND NANOTECHNOLOGY IN ENGINEERING

Vijay K Varadan
University of Arkansas, USA

A Sivathanu Pillai
Debashish Mukherji
Mayank Dwivedi
Defense Research and Development Organization, India

Linfeng Chen
University of Arkansas, USA

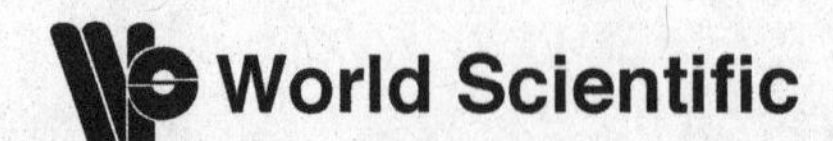

NEW JERSEY • LONDON • SINGAPORE • BEIJING • SHANGHAI • HONG KONG • TAIPEI • CHENNAI

Published by

World Scientific Publishing Co. Pte. Ltd.
5 Toh Tuck Link, Singapore 596224
USA office: 27 Warren Street, Suite 401-402, Hackensack, NJ 07601
UK office: 57 Shelton Street, Covent Garden, London WC2H 9HE

British Library Cataloguing-in-Publication Data
A catalogue record for this book is available from the British Library.

NANOSCIENCE AND NANOTECHNOLOGY IN ENGINEERING

ISBN-13 978-981-4277-92-1
ISBN-10 981-4277-92-4

Printed in Singapore.

Preface

Nanoscience and nanotechnology involve studying and working with materials at the nanometer scale, and stretch across the whole spectrum of science and technology. Generally speaking, nanoscience concentrates on the fundamental relationships between the physical properties of materials and their nanoscale dimensions, and the underlying sciences for nanoscale synthesis, assembly and characterization. Nanotechnology mainly deals with the design, fabrication and application of nanomaterials. The research and development in nanoscience and nanotechnology require the collaborations between researchers from different disciplines, such as physics, chemistry, materials science, electrical engineering, mechanical engineering and biomedical engineering. The ultimate goal of nanoscience and nanotechnology is to develop materials, devices and systems that outperform the ones developed based on the conventional sciences and technologies, and to create completely novel functionalities.

The advances in nanoscience and nanotechnology exhibit great varieties, and the nanoscience and nanotechnology in engineering is one of the most active and the most important areas in the field of nanoscience and nanotechnology. It directly relates the research activities in nanoscience and nanotechnology to industries and daily life, and numerous nanomaterials, nanodevices and nanosystems for various engineering purposes have been developed and applied for human betterment. Many universities are developing and delivering courses on this area, and we therefore realize the urgent need for an appropriate textbook.

This is the first book that systematically discusses the engineering aspects of nanoscience and nanotechnology from the fundamental level, and includes the breakthrough milestones and the latest developments in this area. It is specially tailored to be a textbook for undergraduates, graduates and those seeking short-term professional trainings. It may also serves as a reference desk resource for both academic and industrial researchers interested in this area.

This book consists of eight chapters. Each chapter provides an overview of a specific topic with examples chosen primarily for their educational purposes, and it is organized in a way that fits readers from different science and engineering disciplines. Students are encouraged to expand on the topics discussed in this book by reading the references provided at the end of each chapter.

The first chapter provides the fundamental knowledge of nanoscience and nanotechnology. It starts with introducing the origins of nanoscience and nanotechnology, followed by discussing the general schemes and classification of nanotechnology. The challenges in nanoscience and nanotechnology are analyzed subsequently. Chapter 2 discusses the physical and biological aspects of nanoscience and nanotechnology. After introducing the basics of quantum physics and the fundamentals of nanophysics, the crystal structures and physical properties of materials are discussed, followed by the physical aspects of nanochemistry. After that, the biological aspects of nanoscience and nanotechnology are discussed.

Chapter 3 outlines the technologies for nanoscale fabrication and characterization. These techniques provide the experimental bases for nanoscience and nanotechnology in engineering. This chapter starts with the approaches for nanoscale fabrications, including bottom-up approach and top-down approach. The techniques for characterizing the unique properties of nanomaterials are discussed subsequently, including atomic structure, chemical composition, size, shape and surface area, and the properties of nanoparticles in biological systems.

Based on the fundamental concepts, theoretical backgrounds and experimental techniques discussed in the first three chapters, the following five chapters discuss five important topics on nanoscience and nanotechnology in engineering, including carbon nanomaterials, nanostructured materials, polymer nanotechnology, nanocomposites and organic electronics.

Among different kinds of nanomaterials, carbon nanomaterials are the most popular. Chapter 4 presents a systematic discussion on various types of carbon nanomaterials, including fullerenes, carbon nanotubes and carbon nanofoams. For each type of carbon nanostructures, their synthesizing methods, specific properties and typical applications are discussed.

The properties of nanomaterials are strongly dependent on their structures at the nanometer scale. Chapter 5 deals with the synthesis, properties and applications of typical nanostructured materials, including nanopowders, nanoporous materials, nanodusts, nanowires and nanotubes. Special attention is paid on three-dimensional zinc oxide nanostructures.

Chapter 6 concentrates on a special kind of soft nanotechnology, polymer nanotechnology. After introducing electroactive polymers, the fabrication of polymer nanowires, polymer nanotubes, and three-dimensional polymer nanostructures are discussed.

Due to their unprecedented combinations of properties, nanocomposites are widely used for engineering applications. Various types of nanocomposites are discussed in Chapter 7, mainly including ceramic matrix nanocomposites, metal matrix nanocomposites, magnetic nanocomposites, polymeric nanocomposites, and nano-bio-composites. At the end of this chapter, a brief discussion is made on smart and intelligent nanocomposites, which are regarded as the future of materials science and technology.

The use of individual molecules, such as carbon nanotubes or other organic compounds, as electronic components offers promising alternatives to current microelectronic devices. Chapter 8 deals with nanoscale electronics with a focus on organic electronics. After discussing the fabrication and properties of pentacene thin films, typical organic sensors and their applications are discussed, with emphasis laid on strain sensors and ion-sensitive field effect transistors.

In preparing this book, it is always kept in our mind to relate the speculative concepts in nanoscience and nanotechnology to practical research and development activities. This book contains our many years of experiences and achievements in this area, and a lot of technical details are released for the first time.

Some of the materials for this book are taken from many lectures and courses we presented around the world. The valuable comments from the participants of these lectures and courses greatly enriched the contents of this book. Meanwhile, we would like to indicate that this book is a compilation of the work carried out by many researchers, and we greatly appreciate their valuable contributions in this area. We are also very grateful to the publisher and the staff for their constant encouragement, guidance and support during this project.

There are many people to whom we owe our gratitude for helping us in this process. However, space dictates that only a few of them can receive formal acknowledgements. Our foremost appreciation goes to the research professors and staff of the High Density Electronics Center, University of Arkansas, Fayetteville, in particular, Dr. T. Ji, Dr. H. Yoon and Dr. J. Xie.

In addition, we wish to thank K. Hariharan for helping us in collating the contents of the book and for his constructive suggestions, and Naveen Kalania for his commendable efforts in the design of many diagrams. We would also like to thank the BrahMos Knowledge Centre for providing many inputs for this book.

V. K. Varadan
S. Pillai
D. Mukherji
M. Dwivedi
L. F. Chen

About the Authors

Dr. Vijay K. Varadan is currently the Twenty-First Century Endowed Chair in Nano- and Bio-Technology and Medicine, and Distinguished Professor of Electrical Engineering and Distinguished Professor of Biomedical Engineering (College of Engineering) and Neurosurgery (College of Medicine) at University of Arkansas. He is also a Professor of Neurosurgery at the Pennsylvania State University College of Medicine. He joined the University of Arkansas in January 2005 after serving on the faculty of Cornell University, Ohio State University and Pennsylvania State University for the past 32 years. He is also the Director of the Center of Excellence for Nano-, Micro-, and Neuro-Electronics, Sensors and Systems and the Director of the High Density Electronics Center. He has concentrated on the design and development of various electronic, acoustic and structural composites, smart materials, structures, and devices including sensors, transducers, Microelectromechanical Systems (MEMS), synthesis and large scale fabrication of carbon nanotubes, NanoElectroMechanical Systems (NEMS), microwave, acoustic and ultrasonic wave absorbers and filters. He has developed neurostimulator, wireless microsensors and systems for sensing and control of Parkinson's disease, epilepsy, glucose in the blood and Alzheimer's disease. He is also developing both silicon and organic based wireless sensor systems with RFID for human gait analysis and sleep disorders and various neurological disorders. He is a founder and the Editor-in-Chief of the Journal of Smart Materials and Structures. He is the Editor-in-Chief of the Journal of Nanomedical Science in Engineering and Medicine. He is an Associate Editor of the Journal of Microlithography, Microfabrication and Microsystem. He serves on the editorial board of International Journal of Computational Methods. He has published more than 500 journal papers and 14 books. He has 13 patents pertinent to conducting polymers, smart structures, smart antennas, phase shifters, carbon nanotubes and implantable device for Parkinson's patients, MEMS accelerometers and gyroscopes. He is a fellow of SPIE, ASME, Institute of Physics, Acoustical Society of America. He has many visiting professorship appointments in leading schools overseas.

Dr. Apathukatha Sivathanu Pillai is presently Distinguished Scientist and Chief Controller, Research and Development, DRDO and CEO and MD, BrahMos Aerospace. Electrical Engineer by profession, Dr. Pillai had an opportunity of working with three great aerospace visionaries of India, Dr. Vikram Sarabhai – Architect of India's Space Programme, Prof. Satish Dhawan – Institution Builder and Dr. APJ Abdul Kalam, Missile Man of India, in Space and Defence Technologies. In the Integrated Guided Missile Development Programme (IGMDP) at DRDO, Dr. Pillai was Programme Director and contributed for realizing critical technologies for guided missile systems, through networking of academic institutions, R&D laboratories and industries, leading to successful results and building indigenous capability in many vital systems. His technology leadership capabilities gave him a unique position of CEO and MD of India-Russia Joint Venture BrahMos Aerospace with responsibilities from design, development, production to market the most advanced supersonic cruise missile BRAHMOS, which has been successfully inducted by the Indian Army and Indian Navy. Dr. Pillai is instrumental in progressing nanotechnology research in many academic institutions through nanotechnology courses, evolving syllabus, projects and establishment of centers. His efforts led to the initiation of design and development of many nanotechnology devices. He also contributed in the development of many societal healthcare products as spin-off from the defense technology. He is Honorary Professor of number of institutions and Fellow and Member of many Professional Societies including IEEE, Indian National Academy of Engineering, Instrument Society of India, Astronautical Society of India. Dr. Pillai was awarded with many honorary degrees from Indian and foreign Universities. Dr. Pillai has published many papers in the international and national journals and has three books to his credit. Among the many awards, he is recipient of the Padmashri awarded by the Government of India in recognition of his distinguished contribution in the field of Science and Engineering.

Dr. Debashish Mukherji received his Master of Science in Physics in 1986 and Ph.D degree in Physics – Laser Technology (Fabrication of CO_2 Laser and Appl. to Materials Processing) from Bhopal University, Bhopal, India in 1989. Dr. Mukherji worked for DRDO from 1991 to 2008 as Senior Scientist and contributed in the full system development, which included sub-system level technologies involving Aeronautics, Mechanical and Chemical Engineering, Laser and Optics and Diagnostics and Instrumentation. He has published 14 papers in national and international journals and reports. He is a member of Indian Laser Association, ILA and many awards to his credit.

Mayank Dwivedi is a Post Graduate in Polymer Science and Engineering and is currently working as a Senior Scientist in Defence Research and Development Organisation, Ministry of Defence, India. He is currently pursuing his Ph.D from Indian Institute of Technology, New Delhi, India in the area of Polymer Science and Composites.

Dr. Linfeng Chen received his B. Sc. degree in modern applied physics (major) and his B. Eng. degree in machine design and manufacture (minor) from the Tsinghua University, Beijing, China, in 1991, and he received his Ph.D degree in physics from the National University of Singapore in 2001. From 1991 to 1994, he was an Assistant Lecturer in the Department of Modern Applied Physics, Tsinghua University. From 1994 to 1997, he was a Research Scholar with the Department of Physics, National University of Singapore. In 1997, he joined the Singapore DSO National Laboratories, as a project engineer, and two years later, he became a Member of Technical Staff. From 2001 to 2005, he was a research scientist at the Temasek Laboratories, National University of Singapore. From 2005 to 2009, he was a senior research associate at the High Density Electronics Center, University of Arkansas, Fayetteville. Since November 2009, Dr. Chen has been a research assistant professor at the Arkansas Institute of Nanoscale Materials Science and Engineering, University of Arkansas, Fayetteville. He is senior member of IEEE. His research interests mainly include microwave electronics, electromagnetic functional materials, nanomaterials and nanomedicine.

Dr. [illegible] Mukherjee received his Master of Science [illegible] in 1984 and [illegible] in Physics [illegible] Technology [illegible] Institution of [illegible] Bhopal, India in 1987. Dr. Mukherjee worked for DRDO from 1987 to 1996 as Senior Scientist and [illegible] in the [illegible] which included [illegible] Aeronautical, Mechanical and Chemical Engineering [illegible] and Optics [illegible] and instrument [illegible] published [illegible] national and international journal [illegible] papers. He is a member of [illegible] Laser Association [illegible] and many awards to his credit.

Mayank Dwivedi is a Ph.D. candidate in [illegible] School [illegible] Engineering and is currently working as a Senior Scientist at Defence Research and Development [illegible] Ministry of Defence [illegible] currently pursuing his Ph.D. [illegible] Institute of Technology [illegible] New Delhi [illegible] Science and Computation.

Dr. Lufeng Chen received [illegible] degree in material physics and chemistry [illegible] and [illegible] degree in [illegible] from the Institute [illegible] Beijing, China in 1998 [illegible] from the National University of Singapore in 2003. From 2001 to 2003 [illegible] the Department of Modern Applied Physics [illegible] University from 1994 to 1998 [illegible] Research Scholar with the Department of Physics, National University of Singapore. In 2003, he joined the Singapore [illegible] Institute of [illegible] Later [illegible] he was a research scientist at the [illegible] Laboratory [illegible] University of Singapore. From 2005 to 2007, he was a senior research associate [illegible] Since November 2007, Dr. Chen has been [illegible] associate professor at the [illegible] Institute of [illegible] Science and [illegible] University of [illegible] He is a senior member of [illegible] His research interests mainly include [illegible] functional materials [illegible] devices and [illegible]

Contents

Chapter 1

Introduction

There has been a technology explosion in the last few years due to various new tools and techniques for the creation, characterization and manipulation of materials at the scale of nanometer (nm), which is one billionth of a meter. The technology to create, observe and manipulate matters at the nanometer scale is probably the very field where different disciplines, such as physics, chemistry, materials science and engineering, biology and medicine, combine to offer immense opportunities and challenges. Nanotechnology is the subject, in which paradoxically, you first have to think small to think big and occasionally, make big to make small. It is indeed an ongoing technology revolution. In the coming years, there would hardly be any aspect of our lives where this technology would not make an impact.

The usage of a technology by the society is an indirect measure of advancement and development of the society, and often the nature of the society is known by the technology it excels in. Therefore, the societies have evolved from agriculture based to industry based, and further to information and communications technology based. The period where we stand today is the true manifestation of the increased global knowledge base coupled with the advent of many new technologies.

Technologies such as nano, bio and info would probably make the most profound impact [1-7]. Nanotechnology and biotechnology are expected to influence almost all sectors of the society. Thus, it would bring about a major revolution and influence on the way we live today. The vast global research on these technologies is going to make the world more competitive in terms of both the knowledge and the use of these technologies. As shown in Figure 1.1, we are experiencing the nano-bio revolution and the triad of nano-info-bio technologies, and we will see many new technologies and their products coming into the mainstream.

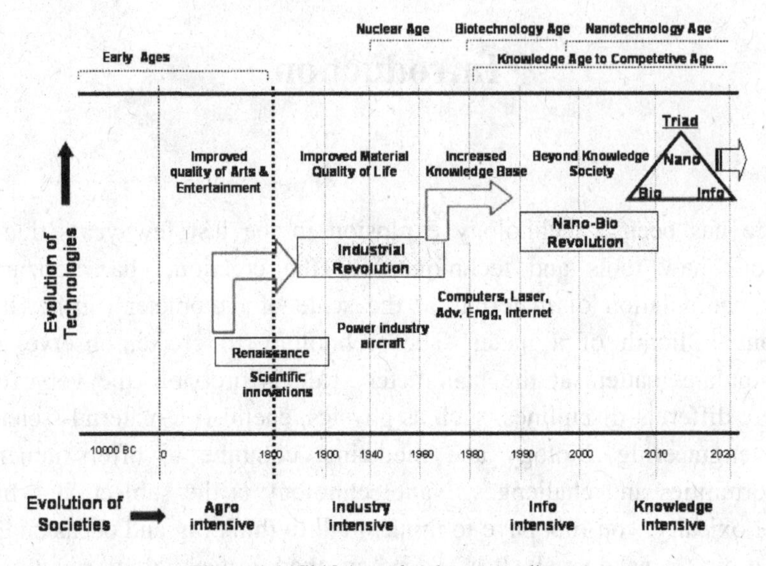

Figure 1.1 Shift in technology paradigm.

1.1 What is Nanoscience and Nanotechnology?

Nanoscience deals with the scientific study of objects with sizes in the 1-100 nm range in at least one dimension. To have an idea about how small the scale is, take a look at the back of your hand. By focusing your naked eyes, you can see down to a scale of a millimeter (one-thousandth of a meter), and the skin looks flat at this scale. However, with the help of a magnifying glass, you will find that the skin is actually wrinkly with cracks and folds. By using a microscope, you could examine the cells that make up your skin. Now you are working at the scale of micrometer, which is one-thousandth of a millimeter. A nanometer (nm) is one-thousandth of a micrometer, or one-billionth of a meter (10^{-9} meter). This is the scale that is often used in measuring atoms and molecules. For example, a hydrogen atom, the smallest elemental particle in nature, has a diameter of about one tenth of a nanometer; six bonded carbon atoms have a width of about one nanometer; and a human hair has a diameter of about 80,000 nm. A typical molecule with the required complexity for nanoscience usually comprises of about 100 atoms and has a diameter in the range of 1 to 10 nanometers.

Generally speaking, nanotechnology is the technology of arranging and processing atoms and molecules for the fabrication of materials with nano specifications [4]. The potential benefits of nanotechnology are both pervasive and revolutionary. With the help of nanotechnology, scientists and engineers can develop materials with remarkable properties, by constructing matters, atom by atom, and molecular by molecular. The matters at nanoscale are quite different from those at macroscale, and due to their special chemical, biological, electrical and magnetic properties, nanoscale matters are becoming the building blocks in creating faster, cheaper and state of the art miniaturized products, as shown in Figure 1.2.

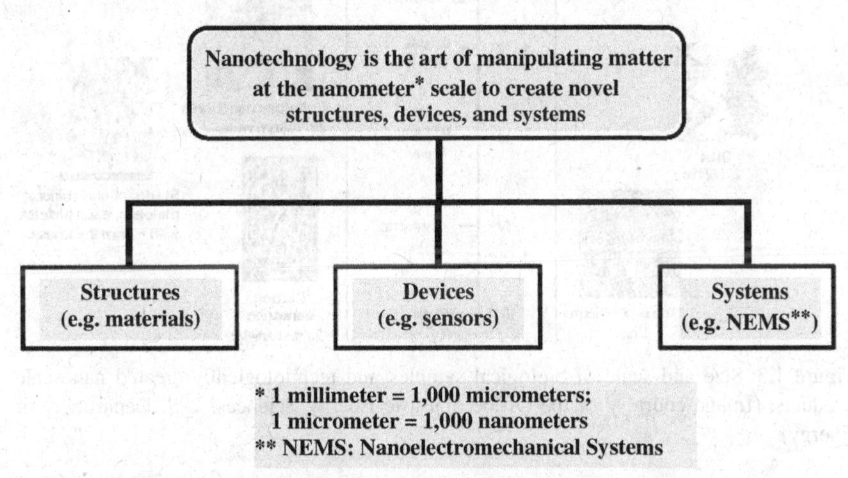

Figure 1.2 Manipulation of matter at nanoscale for materials, sensors and systems.

1.1.1 *Natural and artificial nanoparticles*

Nanoparticles are the simplest nanostructures with dimensions in the nanometer range. In a broad sense, any collections of atoms with a structural radius less than 100 nm can be regarded as nanoparticles, such as fullerenes and proteins. Generally speaking, nanoparticles can be classified into two categories: natural nanoparticles and artificial nanoparticles. Figure 1.3 visualizes various naturally occurring products with regard to technologically created nanoscale products.

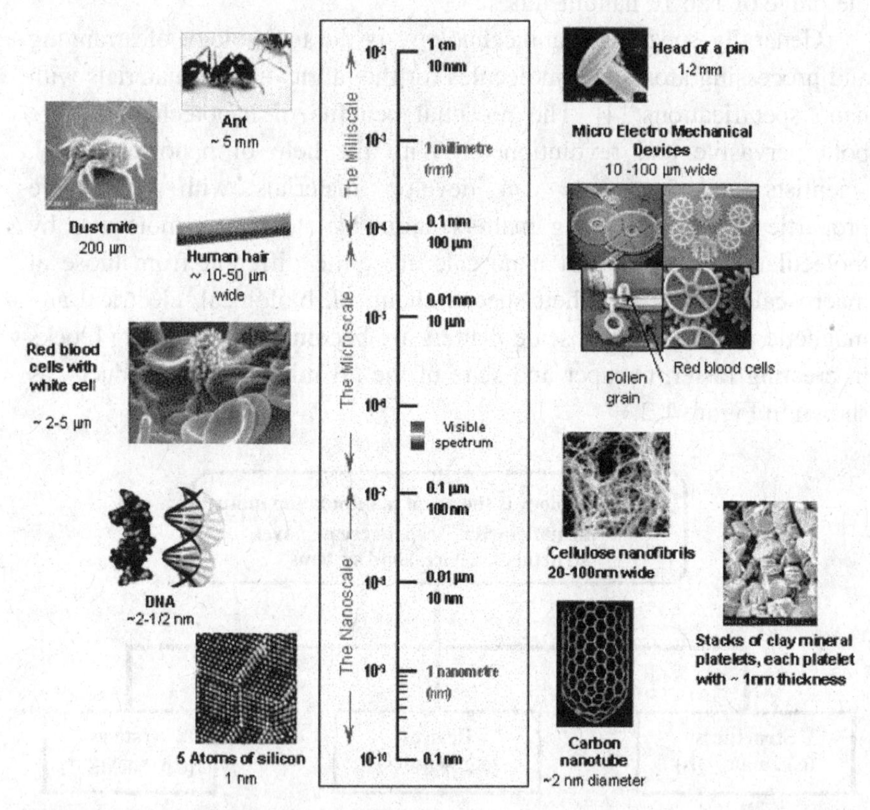

Figure 1.3 Size and scale of biological samples and technologically created nanoscale products. (Image courtesy of the Office of Basic Energy Sciences, U.S. Department of Energy)

The nature is ripe with nanoparticles. For example, when water drops form in clouds or mists, they nucleate from free water molecules, starting their growth in the nanoscale size regime. In fact, many particles, such as water-ammonia, hydrocarbon molecules from car exhaust gases, viruses and bacteria, stay in the nanometer size regime.

Even the lotuses, the icon of purity, could well serve as a symbol of a blooming nanotechnology. W. Barthcott, a German botanist, explained why the lotus leaves stay dry even in a heavy shower. Lotus leaves are coated with hydrophobic wax crystals of about 1-2 nm in diameter which forms the nanostructure with the possible self-cleaning ability. The same technique can be applied to clean wood, paper, textiles, masonry and leather which would result in self-cleaning roofs, shoes and clothes. A German company is developing a spray-on coating that mimics lotuses. The idea of self-cleaning material has been demonstrated in another application. A glass, with a layer of titanium-dioxide based ultra-thin coating (about 40 nm thick), has the capability to clean itself. Unlike the ordinary glass which retains droplets of water, such a glass will spread the rainwater evenly on its surface and let it run off. The organic debris on the glass will be broken down when the ultraviolet waves in sunlight react with a photoresist.

In the study of nanoparticles, a distinction should be made between the "hard" condensed-matter nanoparticles, which are usually called nanoclusters, and "soft" bio-organic nanoparticles and large molecules. A nanocluster is a nanometer-sized structure consisting of subunits, which can be atoms of a single element, molecules or combinations of atoms of several elements with equal stoichiometries.

One distinctive difference between nanoclusters and molecules is that the properties of a nanocluster are solely guided by the properties and number of its subunits, while the functionality of a molecule is mainly determined by the inter-positioning of the atoms. Therefore, nanoclusters can be taken as an intermediate stage between single atoms/molecules and bulk materials. Generally speaking, the physical properties of a material, such as conductivity and elasticity, depend on its dimensions, and are scalable with respect to the amount of atoms in the material. However, as the number of atoms in a nanocluster is very small, every addition or subtraction of an atom to a nanocluster will greatly change the properties of the nanocluster.

1.1.2 *Nanotechnology in ancient time*

Though the word "nanotechnology" is new, nanotechnology itself has existed for a long time. Remarkable understanding of a matter at atomic scale was perceived by an ancient Indian scholar, Kanada (600 BC), and

Democritus (400 BC) of Greece. According to Kanada, atoms compose all things in the world. Democritus coined the word "atom", meaning "not cleavable" in Greek, and held that each matter consisted of atoms and atoms were indivisible. Gold nanoparticles as an inorganic dye to induce red color can be found in Ming dynasty pottery, art pieces in China, and stained glass windows in medieval churches, though the idea of nanoparticles would not have been so well understood by the craftsmen. Figure 1.4 shows the Rayleigh light scattering of nanocrystals with specific shapes, sizes and composition matters [27]. A 100 nm Ag nanocrystal and a 100 nm Au nanocrystal are red and orange respectively. But as they go to the smaller sizes, their colors undergo radical changes.

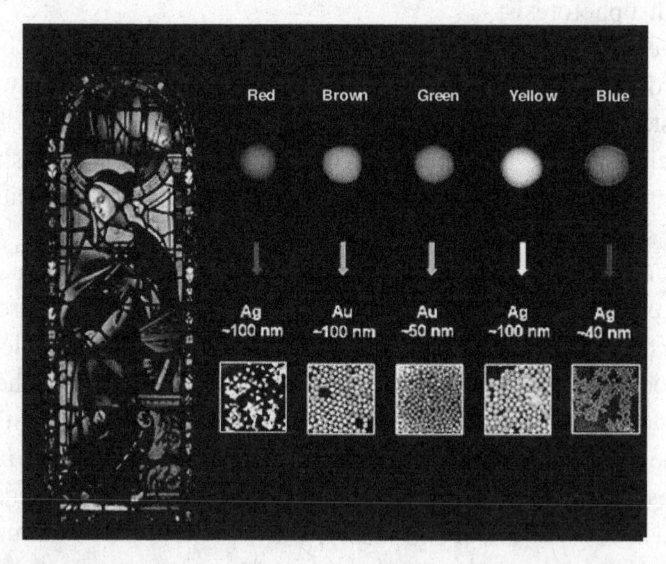

Figure 1.4 Rayleigh light scattering of nanocrystals with specific sizes, shapes and composition matters.

However, it should be indicated that, although the study on materials in the nanometer scale can be traced back for centuries, the current nanotechnology is mainly driven by the ever shrinking of devices in electronics and semiconductor industry, and is supported by the availability of the characterization and manipulation techniques at the nanometer level.

1.1.3 *Feynman, Drexler and Taniguchi*

In the history of nanoscience and nanotechnology, three scientists made great contributions to this field: Richard Feynman, Norio Taniguchi and Eric Drexler. Due to their contributions, we have the modern nanoscience and nanotechnology.

In the mid-1980s, our ability to manipulate objects and make artifacts on this scale of dimensions has brought some remarkable technological developments and promises to bring huge benefits to the 21st century society. Richard Feynman, American physicist and Noble Laurate, presaged the whole field. In 1959, he published his famous paper entitled "*There's plenty of room at the bottom: an invitation to enter a new field of physics*" [8]. In the paper, he encouraged scientists to "think small", and indicated that, with the ability to manipulate matter on very fine, even atomic scales, numerous breakthroughs could be made. He predicted the development of electron beam lithography, ultra-fast integrated circuits, and even the ability to make objects by picking and placing single atoms. Justifiably, he is the father of this field.

However, the credit for coining the term "nanotechnology" must be given to Norio Taniguchi. In his lecture on ultra precision machining in 1974, he used the term nanotechnology. In the same year, the first such device was patented by IBM scientists. In 1981, a young student of MIT, Eric Drexler, published the first technical paper on molecular nanotechnology. Later on in 1983, he presented the complete description of a molecular computer, and there after in 1986, he published the first book on molecular nanotechnology, *Engines of Creation* [9]. If Feynman was the philosopher, Drexler was the prophet. His novel ideas and understanding of the subject earned him the nickname "Mr. Nano".

1.1.4 *Moore's law*

In the past decades, great progress has been made in nanoscience and nanotechnology. As technology kept on developing, many machines and components kept on reducing. Meters long machines became centimeters long and even smaller. As shown in Figure 1.5, according to the Moore's Law, the number of transistors in successive generations of computer chips has risen exponentially, doubling every 1.5 years or so [10]. The notation "mips" on right ordinate is "million instructions per second". Gordon Moore, co-founder of Intel Inc., predicted this growth pattern in 1965, when a silicon chip contained only 30 transistors. The number of dynamic random access memory (DRAM) cells follows a similar growth pattern. The growth is largely due to continuing reduction in the size of the key elements in the devices, to about 100 nm, with improvements in optical photolithography.

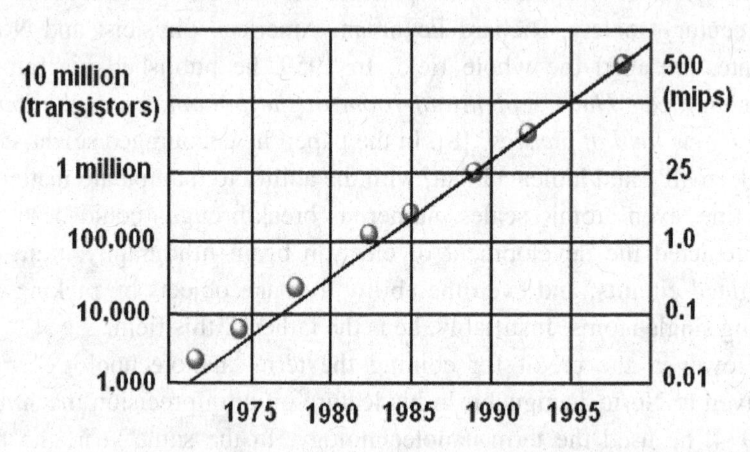

Figure 1.5 Moore's Law: the growing power of silicon chips, measured in millions of instructions per second [10].

By extending the trends shown in Figure 1.5, we are coming to the nanometer level, which is the size of atoms and molecules. The investigation of the fundamental facts of atoms and molecules for understanding and exploring the applications of nanoscience and nanotechnology is an important topic at this level.

1.2 Effects of Making into Small

To appreciate and explore the applications of nanoscience and nanotechnology, it is necessary to understand why and how the nanometer scale affects materials properties [1, 4]. The effects of the nanometer scale mainly include:

- The small length scales present in nanoscale systems directly influence the structures of energy bands, and indirectly lead to the changes in the associated atomic structure. These effects are generally termed quantum confinement, which changes the total energy of a system.
- Reduction of system size may change the chemical reactivity of the system, which is a function of the structure and occupation of the outermost electronic energy levels. Furthermore, the physical properties such as electrical, thermal, optical and magnetic characteristics, which also depend on the arrangement of the outermost electronic energy levels, can be changed due to size reduction. For example, metallic systems can undergo metal-insulator transitions as the system size decreases, due to the formation of a forbidden energy bandgap.
- The mechanical strength, which depends on electronic structure, may be affected by size reduction. In addition, owing to the change of the electron energy levels, the transport properties of a material at nanometer scale may exhibit quantized rather than continuous behaviors.

The changes of the properties of nanostructures caused by reduced dimensions are related to the proportion of atoms which are in contact with either a free surface, as in the case of isolated nanoparticles, or an internal interface, such as a grain boundary in a nanocrystalline solid. If an atom is located at a surface, the number of its nearest-neighbor atoms is reduced, giving rise to differences in bonding (leading to the well-known phenomenon of surface tension or surface energy) and electronic structure. In an isolated nanoparticle, a large proportion of the total number of atoms will be present either at or near the free surface. For instance, in a 5 nm particle approximately 30-50% of the atoms are influenced by the surface, compared with approximately a few percent for a 100 nm particle. Similar arguments apply to nanocrystalline materials, where a large proportion of atoms will be either at or near grain boundaries. Such structural differences in reduced-dimension systems would be expected to lead to very different properties from those of bulk materials.

1.2.1 *Size dependence of materials properties*

The way how the nanometer scale affects the properties of materials are related to nanophysics and nanochemistry, which are discussed in Chapter 2. As an example, here we discuss how the yield strength is related to the grain size.

The yield strength σ_y is a standard engineering parameter indicating how much a material can be drawn out before it fails for practical purposes. The yield strength of a polycrystalline material follows the Hall-Petch relationship given by:

$$\sigma_y = \sigma_0 + \left(K / \sqrt{d}\right) \quad (1.1)$$

where σ_y is the yield strength, d is the mean grain size, and σ_0 and K are constants. The explanation for Hall-Petch relationship is rather complex but related to dislocation activity in the grains. According to Eq. (1.1), if the grain size of a material is decreased, its yield strength will be increased. However, atomistic simulations of deformation indicate that materials with very small grain sizes exhibit a reverse Hall-Petch effect: the yield strength of a material decreases with the decrease of its grain size. Usually the reverse Hall-Petch effect happens when the grain size is less than 20 nm.

It should be indicated that many properties of a material are related to the size of the material, and usually they are extrinsic properties. For example, the resistance of a specimen depends on the shape and size of the specimen. While the intrinsic properties of a material, such as resistivity, are independent of the shape and size of a specimen. However, at the nanoscale, many of the intrinsic properties of a matter are not necessarily predictable from their corresponding ones at larger scales. The reason for this is that at the nanometer scale, totally new phenomena may emerge, such as quantum confinement resulting in the changes in electronic structure, the presence of wave-like transport processes, and the predominance of interfacial effects.

1.2.2 *Special properties of nanomaterials*

Usually, materials at the micrometer scale exhibit the same physical properties as those of bulk materials. While at the nanometer scale, materials may exhibit quite different physical properties from those of bulk materials. As mentioned above, when the dimension of a material approaches the nanoscale, new properties emerge due to size confinement, quantum phenomenon and Coulomb blockage. The transition from atomic or molecular to the bulk form takes place in this range. Crystals in the nanometer scale have low melting point along with reduced lattice constants. This primarily occurs as the number of surface atoms or ions and surface energy plays a predominant role in thermal stability. Ferroelectrics and ferromagnetics may lose their ferroelectricity and ferromagnetism when they shrink to nanometer dimensions, and bulk semiconductors become insulators. Using nanotechnologies, the properties of materials can be controlled for special applications, and a great deal of research work in nanotechnology aims for the control of structural properties, thermal properties, chemical properties, mechanical properties, magnetic properties, optical properties, electronic properties and biological properties [11-13, 26].

1.2.2.1 *Structural properties*

The increase in surface area and surface free energy with the decrease of particle size leads to the changes in interatomic spacing. For example, the interatomic spacing of a Cu metallic cluster decreases with the decrease of the cluster size. This effect is mainly due to the compressive strain induced by the internal pressure arising from the small radius of curvature in the nanoparticle. Conversely, for semiconductors and metal oxides, their interatomic spacing increases with the decrease of the particle size.

1.2.2.2 *Thermal properties*

The increase in surface energy and the change in interatomic spacing have strong effects on the physical properties of a material. For instance, it has been observed that the melting point of gold particles, which is a bulk thermodynamic characteristic, decreases rapidly for particles with sizes less than 10 nm. However, the opposite behavior can be observed for metallic nanocrystals embedded in a continuous matrix: smaller particles have higher melting points.

1.2.2.3 *Chemical properties*

The change in structure as a function of particle size is intrinsically related to the changes in electronic properties. Generally speaking, the ionization potential, the energy required to remove an electron, for small atomic clusters is higher than that for the corresponding bulk material. Meanwhile, it is found that the ionization potential may exhibit obvious fluctuations as a function of the cluster size. It appears that such effects are linked to chemical reactivity, such as the reaction of Fen clusters with hydrogen gas.

1.2.2.4 *Mechanical properties*

Many mechanical properties, such as toughness, are strongly dependent on the ease of formation or the presence of defects within a material. When the system size decreases, the ability to support such defects becomes increasingly more difficult and subsequently the mechanical properties will be altered significantly. Nanostructures, which are quite different from bulk structures in terms of atomic structural arrangement, obviously show very different mechanical properties. For example, single- and multi-walled carbon nanotubes exhibit high mechanical strengths and high elastic limits that lead to excellent mechanical flexibility.

In general, metals have high ductility, but they are not elastically hard. Ceramics are usually elastically hard, but they have very low ductility. However, when the grain sizes of these materials are in the nanometer scale, metals become elastically harder and less ductile, while ceramics have higher ductility. Therefore, from a mechanical point of view, nanotechnology provides approaches for optimizing the properties of metals and ceramics.

1.2.2.5 *Magnetic properties*

Magnetic nanoparticles have extensive applications, such as ferrofluids, color imaging, bioprocessing, refrigeration as well as high storage density magnetic memory media. The large surface area to volume ratio results in a substantial proportion of atoms at the surface, which have a different local environment and thus a different magnetic coupling with neighboring atoms, leading to different magnetic properties.

Usually bulk ferromagnetic materials have multiple-domain structures, while small magnetic nanoparticles often consist of only one domain and exhibit superparamagnetism. Superparamagnetic nanoparticles have very low coercivity. Due to thermal fluctuations, the magnetizations of nanoparticles are randomly distributed, and only become aligned in the presence of an applied magnetic field.

1.2.2.6 *Optical properties*

In nanoclusters, the reduced dimensionality on electronic structure has profound effects on the energies of the highest occupied molecular orbital (HOMO), essentially the valence band, and the lowest unoccupied molecular orbital (LUMO), essentially the conduction band. Optical emission and absorption depend on the transitions between these states. In particular, semiconductors and metals show large changes in optical properties, such as color, as a function of particle size. For example, colloidal solutions of gold nanoparticles have a deep red color, and the color becomes progressively more yellow as the particle size increases. Since the 17th century, gold colloids have been used as a pigment for stained glass.

1.2.2.7 *Electronic properties*

The changes in electronic properties as the length scale is reduced are related mainly to the increasing influence of the wave-like property of the electrons and the scarcity of scattering centers. As the size of the system becomes comparable with the de Broglie wavelength of the electrons, the discrete nature of the energy states becomes apparent, although a fully discrete energy spectrum could only be observed in systems that are confined in all three dimensions.

In certain cases, conducting materials become insulators below a critical length scale, as the energy bands cease to overlap. Owing to their intrinsic wave-like nature, electrons can tunnel quantum mechanically between two closely adjacent nanostructures. If a voltage is applied between two nanostructures which aligns the discrete energy levels in the density of states, the tunneling current will abruptly increase due to the resonant tunneling effect.

For a macroscopic system, the electronic transport properties are primarily determined by the scatterings with phonons, impurities and other carriers, or by the scatterings at rough interfaces. Usually, the path of each electron resembles a random walk, and the transport is diffusive. When the dimensions of a system are smaller than the electron mean free path for inelastic scattering, electrons can travel through the system without randomization of the phase of their wave-functions. This gives rise to additional localization phenomena that are specifically related to phase interference. If the system is sufficiently small so that all the scattering centers can be completely eliminated, and the sample boundaries are sufficiently smooth so that boundary reflections are purely specular, the electron transport becomes purely ballistic, and the sample acts as a waveguide for the electron-wave function.

Conduction in highly confined structures, such as quantum dots, is extremely sensitive to the presence of other charge carriers, so the charge state of quantum dots is important. These Coulomb blockade effects result in conduction processes which involve single electrons, and as a result they require only a small amount of energy to operate a switch, a transistor or a memory element.

The phenomena discussed above could be utilized to develop original types of components for applications in electronics, opto-electronics and information processing, such as resonant tunneling transistors and single-electron transistors.

1.2.2.8 *Biological properties*

Biotechnology is a form of nanotechnology also known as the wet side of nanotechnology. All thc nanomachines of cellular life and viruses are grouped in this category.

Biological systems contain many examples of nanophase materials and nanoscale systems. Biomineralization of nanocrystallites in a protein matrix is highly important for the formation of bones and teeth, and is also used for chemical storage and transport mechanisms within organs. Biomineralization involves the operation of delicate biological control mechanisms for the synthesis of materials with well-defined characteristics such as particle size, crystallographic structure, morphology and architecture.

Complex biological molecules, such as DNA, usually have the ability to undergo highly controlled and hierarchical self-assembly, which makes them ideal for the assembling of nanosized building blocks. Methods for altering and controlling these nanoscale building blocks and assembling nanoscale architectures are important in biology.

Biological cells have dimensions usually in the range of 1-10 μm, and contain numerous extremely complex nanoassemblies, such as molecular motors, which are complexes embedded within membranes and powered by natural biochemical processes.

Naturally occurring biological nanomaterials have been refined over a long timescale and therefore are highly optimized. We can use biological systems as a guide for producing synthetic nanomaterials and nanosystems, and such a process is often called biomimicry.

1.3 General Themes and Classification of Nanotechnology

1.3.1 *General themes of nanotechnology*

Because nanotechnology is defined by the size of the materials or structures to be investigated, to be developed or to be utilized, the products resulted from the research and development activities in this area, such as fabrics, fuel cells and drug delivery devices, exhibit great diversities. What brings them together is the natural convergence of basic sciences, such as physics, chemistry, biology and medicine, at the molecular level. In the following, we discuss the general themes of nanotechnology by which the diverse fields in nanotechnology are unified.

1.3.1.1 *Characterization tools*

To observe and examine the nanostructures or the building blocks of nanomaterials, characterization tools such as X-ray diffraction (XRD), scanning electron microscopy (SEM), transmission electron microscopy (TEM), scanning tunneling microscopy (STM) and atomic force microscopy (AFM) are indispensable.

1.3.1.2 *Nanoscale science*

Because the properties of materials change in unexpected ways at the nanometer scale, the sciences studying the behaviors of molecules at this scale are critical to the rational design and control of nanostructures for practical applications.

1.3.1.3 *Molecular level computation*

Computation technologies such as quantum mechanical calculations, molecular simulations and statistical mechanics are essential for understanding nanoscale phenomena and molecular interactions.

1.3.1.4 *Fabrication and processing technology*

Though many nanoparticles can be directly applied in paints, cosmetics and therapeutics, a lot of nanomaterials must be assembled and fabricated into components and devices to perform desired functions. In addition, processing techniques, such as sol-gel, chemical vapor deposition, hydrothermal treatment and milling are often used for synthesizing various types of nanoparticles.

1.3.2 *Classification of nanotechnology*

To investigate various fields in the nanoworld, different nanotechnologies have been developed. These technologies generally fall into three categories: wet nanotechnology, dry nanotechnology and computational nanotechnology.

Wet nanotechnology deals with biological systems primarily existing in water environments. The nanoscale objects often investigated in wet nanotechnology mainly include enzymes, membranes, DNA and other cellular components.

Dry nanotechnology is developed from physical chemistry and surface sciences, and it concentrates on the synthesis and characterization of nanostructures in dry environments, such as carbon nanostructures and silicon nanostructures.

In the computational nanotechnology, the analytical and predictive power of computation is used to model and simulate complicated nanoscale materials and structures.

1.4 Challenges in Nanoscience and Nanotechnology

Nanotechnology may be the ultimate enabling technology, since it deals with the fundamental building blocks of matters and lives. Almost every field of industry will be deeply affected by the progress in nanotechnology. The most important impact of this nanotechnology revolution may be the new synergy among scientists, engineers, industrialists, entrepreneurs, financiers and economic development specialists. The nanotechnology creates both challenges and opportunities [14-16].

The major challenges of nanotechnology mainly include technology challenge, societal and ethical challenge, business driven economy, risk and uncertainty, and security and military implications.

1.4.1 *Technological challenge*

In nanoscience and nanotechnology, one major technological challenge is the nanomaterial science required for local growth of nanostructures with desired solid-state or molecular properties, and for the control of local reactions. The second technological challenge concerns nano-interfaces as connections and active components. Another technological challenge deals with the novel components with electronic, mechanical and chemical functionalities, which can be used for energy and information transfer to autonomous nanosystems [17]. It is expected that nanomechanics and nanochemistry will provide original paths between the "virtual" world of all kinds of data processing, including thermal, mechanical and chemical processing, and the "real" world of sensing and actuation. Such paths are crucial for the development of integrated sensing, processing and actuation technologies.

Nominally, nanotechnology involves the manipulation of atoms and molecules. Though, like plants and animals, we could manipulate atoms and molecules using biological processes, we could not freely and accurately control biological processes to perform desired functions. Using our current knowledge and techniques, some biological processes, such as tissue growth, can be guided on a macroscale; however, we could not yet guide these biological processes in the way viruses and enzymes do.

In the research of nanoscience and nanotechnology, one of the most important challenges would be the growth, characterization and functionalization of nanomaterials and nanostructures. It is still a daunting task to control the material property and make an ideal metal, semiconductor or insulator with perceived properties. Unlike other well-known aspects of manufacturing, nanodesign and simulation issues are just beginning to come out from the realms of human imagination.

1.4.2 *Societal and ethical challenge*

Public health and safety is probably the major concern associated with nanotechnology [18]. It is becoming important and urgent to address the societal and ethical issues related to this emerging area of technology. It is necessary to make sure that the fear of the public about nanotechnology be eliminated, and the products of nanotechnology be accepted by the society and are not blocked by the public due to fear.

The possible threats of nanotechnology are related to how nanoparticles affect the environment, and more essentially, how nanoparticles interact with the human body. A lot of efforts and resources have been spent in incorporating nanoparticles into products that have already been marketed to the public, and accepted by the public. However, the research on the health issues related to nanotechnology is very limited, and this is a matter of severe concern.

The elements involved in nanotechnology behave quite differently from their counterparts in conventional technology. For example, the properties of graphite have been extensively investigated and well known, and so there are mature toxicology guidelines for graphite. However, such guidelines cannot be applied to carbon nanomaterials. Though fullerenes and nanotubes could be legally categorized as graphite, they behave quite differently from graphite. Experiments have shown that fullerenes might cause brain damage to fishes and change gene markers in their livers. Therefore, their entire physiology may be affected by fullerenes. Experiments also indicated that water fleas, an important link in the marine food chain, may be killed by fullerenes.

Though it is not quite sure whether fullerenes can cause brain damage to humans, detailed and systematic investigations on this topic are definitely needed. Before we have more knowledge on this topic, it is necessary to avoid the accumulation of fullerenes over time, and especially we should not let them enter the food chain. Experiments have showed that nanoparticles may accumulate in the bodies of lab animals, and fullerenes could travel freely through soil and could be absorbed by earthworms.

The adverse effects of other nanoparticles have also been observed. Experiments have shown that the nanoparticles used in sunscreen might create free radicals that could damage DNA, cadmium selenide nanoparticles could cause cadmium poisoning in humans, and gold nanoparticles might move through a mother's placenta to the fetus.

As nanoparticles with different sizes and shapes have different levels of toxicity, it is difficult to have a uniform category even for a single element. Particles with smaller sizes usually have higher bioactivity and toxicity. As smaller particles can more easily cross the skin, the lung and the blood/brain barriers, they have higher capability to affect living systems. Once nanoparticles go inside the body, they may trigger many biochemical reactions, for example, the creation of free radicals that may damage cells. Though the body has its natural defense system, it seems that the natural defense system is not so effective for artificial nanoparticles, which are totally new for the natural defense system in the body.

The workers producing nanoparticles or products containing nanoparticles are in the highest risk [19]. According to the National Institute for Occupational Safety and Health (NIOSH), there are more than two million Americans exposed to high levels of nanoparticles, and it is believed that in the near future this number may increase to about four million. NIOSH has published safety guidelines and other information for the workers in nanotechnology industry. It seems that we are trying to learn the lesson from asbestos dust. Asbestos dust was regarded to be safe until it was found that it could cause cancer from accumulation in the body. Due to decades-old use of asbestos, even today three thousand deaths per year are due to asbestos. Because of the extensive applications of nanoparticles in various products, such as make-up, car paint and tennis rackets, it is becoming more and more important to figure out methods to avoid the possible dangers of nanotechnology.

1.4.3 *Business driven economy*

Economic benefits are another major challenge. Every technology evolution may bring about great economic benefits. For example, it is expected that the Internet would create millions of new jobs and trillions of dollars of economic activity. However, it should be indicated that it is just a reallocation of the same old money, and the market capitalization mainly comes from manufacturing and resource companies. In a simplistic view, there is no new money, only old money transferred. What comes into one industry must have come from another.

Nanotechnology is the same case. Many new companies based on nanotechnology will be set up, and their stock prices will appreciate at the cost of other companies. Nanotechnology products will earn revenues, by taking the money from the revenues previously accorded to old products, for example, old metals, under-performing drugs, and clothes that are not stain resistant. Old industries will not simply go away, and they would take all possible measures to conserve their markets. Finally, the new technology has to come in at a significant discount to the old [20-24].

1.4.4 *Risk and uncertainty*

Nanotechnology could well pave the way for the next revolution to take place. It is often remarked that it is easier to make destructive use of a technology rather than use it constructively. The risk and uncertainty of nanotechnology should be taken very seriously by both the developed world and the developing world. Nanotechnology could enable new weapons in more than one way.

First, using nanotechnology, existing drugs could be delivered more effectively. Many pharmaceutical companies are working on nanoparticles that could make drugs effectively absorbed by the body. Nanotechnology could be used to develop viable medicines, and meanwhile it can also be used to boost the power of compounds for the development of chemical weapons. For example, delivering nicotine in lethal amounts is quite difficult due to the barriers of the body. Maybe using nanotechnology, something can be developed that could assist nicotine to get through the barriers of the body. In this way, weapons can be made based on something that is not lethal.

Second, due to the rapid progress in biology, it is possible to use nanotechnology to develop new weapons that are not recognized by existing weapons inspectors. For example, since there are well-established means for the synthesis of cyanide, it could be detected at a very early stage by following the production process. However, using nanotechnology, new agents could be developed to attack very specific functions in the body, for example, the central nervous system. As the amount of agent needed is usually quite small, its synthesis does not require big industrial bases, and tracing such agents is very difficult. Some of the above issues have been outlined by Pardo-Guerra, at the College of Mexico, in the *Nanotechnology Law and Business Journal* [25].

1.4.5 *Security and military implications*

Nanotechnology provides a wide range of possibilities for military purposes, which may greatly improve the existing systems and define radically new systems. For example, guns developed using three-dimensional assembly of nanostructures, could be lighter and carry more bullets. Such guns can be so smart that they can find enemies and automatically shoot themselves. In a more revolutionary approach, nanobots and bionanobots might be developed for military applications. By attacking certain kinds of metals, rubbers and lubricants of conventional weapons, specially designed nanobots could destroy these weapons by consuming them. When bionanobots are ingested from the air by a person, they could assay his DNA codes, release specific gene drugs and self-destruct in the body.

Some technologists have warned that nanotechnology would make it possible for terrorists to acquire nuclear weapons with very little fissionable material. The production of such weapons would not require big uranium or plutonium plants, as very small quantities of deuterium-tritium mixture would be adequate. They can be precisely aimed without any elaborate launch facility, and radioactive fallouts can be avoided. The key techniques for such weapons would be available on the Internet, empowering individuals and small groups of terrorists with adequate knowledge.

It deserves to be noted that nanotechnology is a value-neutral tool. Depending on how it is used, nanotechnology can be both good and bad. It is widely accepted that if there is a new role for the military to prevent conflict, not just delivering force, nanotechnology offers great opportunities for a new, proactive approach to national security.

1.4.6 *Emerging concern from nanoparticles*

It is concerned that nanomaterials may prove to be imminent dangers to human health and environment. Though many materials are harmless in their bulk form, when they are reduced to fine nanoparticles, they may be harmful. It is feared that even harmless compounds may turn out to be risky at the nanometer scale. For example, inhaling nanoparticles during their production procedure may harm lungs. In the experiments by the scientists at the NASA's Johnson Space Center in Houston, the mice exposed to carbon nanofibers developed lesions in their lungs and intestines.

The action group on Erosion Technology and Concentration (ETC), Canada, points out that there are no rules to regulate the use of nanoparticles in the laboratory or outside. ETC says synthetic nanoparticles could be toxic, and has called for a worldwide ban on nanotechnology research. Some organizations have started the research on the impact of nanoparticles on the human body.

In a report, the Greenpeace, UK, has conceded that nanotechnology may not be after all harmful and therefore has not demanded a ban on its use. However, the report has clearly voiced the concern about the lack of regulation of nanoparticles. Though valuable benefits may flow from the applications of nanotechnology, such as clean energy, the Greenpeace hopes scientists to be proactive and address the concerns of the common people and frankly pinpoint the risks, if any, in the applications of this technology.

The US Congress has called for more funds to support the research and development regarding the social, ethical and environmental impacts of nanotechnology. Many people suggested caution in using nanoparticles. The concern is all the more because of the exciting possibilities of using nanoscale devices inside the body for imaging and drug delivery, though researchers would have to satisfy themselves that there would not be any adverse reactions on the healthy tissues in experiments.

Besides the health risk questions, the ethical aspect of this new technology has assumed importance [19]. Many critical questions should be answered, among which, two important issues would be ascertained: whether it is right and safe to develop nanotechnology without fully understanding its implications to human health, and how nanotechnology can be kept away from terrorists who can easily acquire it.

References:

1. Wilson, M. (2004). *Nanotechnology, Basic Science and Emerging Technologies*, Chapman and Hall, London.

2. Whatmore, R. W. (2001). Nanotechnology: Big prospects for small engineering, *Ingenia*, 28-33 (February).
3. Rathjen, D. and Read, L. (2005). *Nanotechnlogy: Enabling Technologies for Australian Innovative Industries*, Prime Minister's Science, Engineering and Innovation Council.
4. Cao, G. (2004). *Nanostructures and Nanomaterials – Synthesis, Properties and Applications*, Imperial College Press, London.
5. Marburger, J. H. and Kvamme, E. F. (2005). *Initiative at Five Years: Assessment and Recommendations of the National Nanotechnology Advisory Panel*, President's Council of Advisors on Science and Technology.
6. Roco, M. C., Trew, R. and Murday, J. S. (1999). *Nanotechnology Research Directions*: *IWGN Workshop Report*, National Science and Technology Council.
7. Roco, M. C., Kalil, T. A., Trew, R. and Murday, J. S. (1999). *Nanostructure Science and Technology: A Worldwide Study*, National Science and Technology Council.
8. Feynman, R. (1960). There's plenty of room at the bottom, *Engineering and Science*, **23**, 22-36.
9. Drexler, K. E. (1986). *Engines of Creation*, Anchor Books.
10. Ball, P. (2000). Chemistry meets computing, *Nature*, **406**, 118-120.
11. Iijima, S. (1991). Helical microtubules of graphic carbon, *Nature*, **354**, 56-58.
12. Iijima, S. and Ichihashi, T. (1993). Single-shell carbon nanotubes of 1-nm diameter, *Nature*, **363**, 603-605.
13. Bethune, D. S., Kiang, C. H., Devries, M. S., Gorman, G., Savoy, R., Vazouez, J. and Beyers, R. (1993). Cobalt-catalyzed growth of carbon nanotubes with single-atomic-layer walls, *Nature*, **363**, 605-607.
14. Bond, P. J. (2004). *Speech on Challenges to Nanotechnology Development and Commercialization, Delivered to the National Nanotechnology Initiative 2004*, Department of Commerce, Washington, DC.
15. Arnall, A. H. (2003). *Future Technologies, Today's Choices – Nanotechnology, Artificial Intelligence and Robotics; A technical, political and institutional map of emerging technologies.* Greenpeace Environmental Trust, London.
16. Cygnus Business Consultancy and Research (2006). *Emerging Technologies in India*, Cygnus Business Consultancy and Research.
17. Baltes, H., Brand, O., Fedder, G. K., Hierold, C., Korvink, J. G. and Tabata, O. (2005). *Enabling Technologies for MEMS and Nanodevices*, Wiley-VCH, Weinheim.
18. UK Department for Business, Enterprise and Regulatory Reform. http://www.dti.gov.uk/nanotechnology.
19. Schulte, P. A. and Salamanca-Buentello, F. (2007). Ethical and scientific issues of nanotechnology in the workplace, *Environmental Health Perspectives*, **115**, 5-12.
20. Micro and Nanotechnology Commercialization Education Foundation (MANCEF) (2004). *MANCEF International Micro/Nano Roadmap*, 2nd Edition, MANCEF.
21. Lux Research, (2004). *Sizing Nanotechnology's Value Chain*, Lux Research.

22. Research and Markets, *Nanotechnology in Asia Pacific*, http://www.researchandmarkets.com/reports/c11879.
23. Australian Government (2005). *Australian Nanotechnology: Capability and Commercial Potential*, 2nd Edition, Commonwealth of Australia.
24. Lux Research (2004). *The Nanotech Report: Investment Overview and Market Research for Nanotechnology*, 3rd Edition, Lux Research.
25. Choi, C. Q. (2005). Nano world: nano could lead to new WMDs, Space Daily, http://www.spacedaily.com/news/nanotech-05zza.html.
26. Kelsall, R., Hamley, I. W. and Geoghegan, M. (2005). *Nanoscale Science and Technology*, Wiley.
27. Mirkin, C.A. http://chemgroups.northwestern.edu/mirkingroup/BioNanomaterials2003rev1.htm.

Chapter 2

Physical and Biological Aspects of Nanoscience and Nanotechnology

To investigate the nanoscale phenomena and to appreciate and explore the applications of nanomaterials, it is necessary to understand the physical and biological aspects of nanoscience and nanotechnology [1-3]. In this chapter, we discuss basics of quantum physics, fundamentals of nanophysics, crystal nanostructures of nanomaterials, physical aspects of nanochemistry [4-6]. As nanotechnology has great potential to be used in biology, and meanwhile biology provides many inspirations for synthesis of nanomaterials, in the final part of this chapter, the biological aspects of nanotechnology are discussed.

2.1 Basics of Quantum Physics

In quantum physics, particles are not considered point-like, but instead have wave-like properties [7-10]. In a 1+1 dimension (x, t), a particle is described by a wave function Ψ = (x, t), and the probability of the particle to be obtained in a given space and time P(x, t) is given by:

$$P(x,t)=|\Psi(x,t)|^2\,dx \tag{2.1}$$

In Eq. (2.1), the Ψ function is normalized as:

$$\int_{-\infty}^{\infty}\Psi^*(x,t)\Psi(x,t)dx=1 \tag{2.2}$$

The time–dependent Schrödinger equation is defined as:

$$-\frac{\hbar^2}{2m}\frac{\partial^2\Psi(x,t)}{\partial x^2}+V(x,t)\Psi(x,t)=i\hbar\frac{\partial\Psi(x,t)}{\partial t} \tag{2.3}$$

where V is the potential energy function. If V is independent of time, one can separate the position and time dependence as follows:

$$\Psi(x,t) = \psi(x)e^{-iEt/\hbar} \tag{2.4}$$

By combining Eqs. (2.3) and (2.4), we have:

$$-\frac{\hbar^2}{2m}\frac{\partial^2\psi(x,t)}{\partial x^2} + V(x,t)\psi(x,t) = E\psi(x) \tag{2.5}$$

By using the Hamiltonian operator H defined by:

$$H \equiv -\frac{\hbar^2}{2m}\frac{\partial^2}{\partial x^2} + V(x,t) \tag{2.6}$$

Equation (2.5) can be rewritten as:

$$H\psi(x) = E\psi(x) \tag{2.7}$$

where ψ(x) is an eigen function and E is an eigen value of H. This equation can have different solutions with different energies E_n.

2.2 Fundamentals of Nanophysics

Nanophysics uses quantum mechanics concepts to describe the fundamental issues related to nanomaterials [11]. Different physical values are quantized, leading to novel effects. Downscaling of a classical bulk material into nanoscale can lead to dramatic changes in the behaviors of the material. In the nanorealm, quantum physics comes into play and leads to completely new kinds of behaviors.

Classic physical models assume the continuity of quantities and involve no restrictions concerning very small physical structures. The quantum theory shows that the values of some measurable variables of a system can attain only certain discrete values. The smallest possible jumps in the values of these observables are called quanta. A quantum is the smallest possible unit that has to fit into a nanostructure. It becomes necessary to use the quantum mechanics concepts to explain the operational principles of devices with nanometer structures. In fact, quantum effects set up the limit of electronic miniaturization and information processing.

2.2.1 *Electron levels for atoms and solids*

2.2.1.1 *Atoms*

When the time-independent Schrödinger equation is applied to an individual atom, the potential function $V(r)$ contains the Coulomb interactions between negative electrons and the central nucleus, as well as the interactions between the electrons. Qualitatively this leads to a potential well for electrons as shown in Figure 2.1.

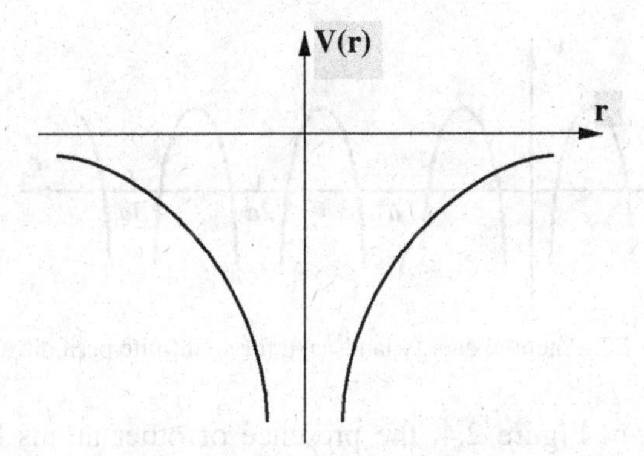

Figure 2.1 Potential well for electrons.

The solutions to this are a set of bound discrete energy levels for the electrons, which then lead to the properties of atoms such as X-ray emission and absorption at given energy levels and so on. This is the basis of atomic physics. But if one takes into account a large collection of atoms, the potential energy landscape for electrons changes as shown in Figure 2.2.

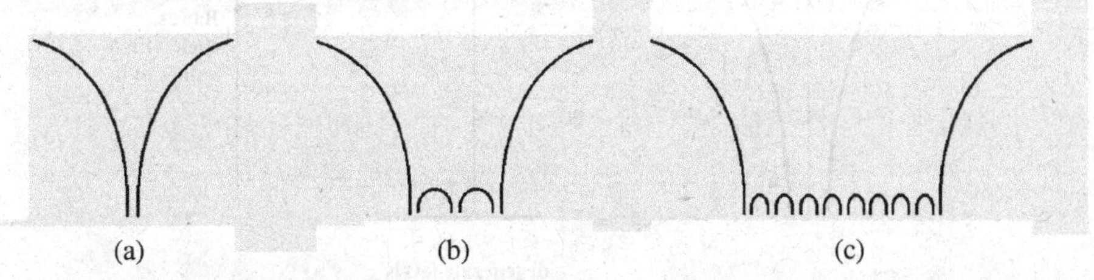

Figure 2.2 Potential energy landscape for electrons. (a) One ion, (b) three ions in line, (c) many closely spaced ions in line.

For a large collection of atoms, the Coulombic potential is valid only at the edges, while there are maxima inside the potential energy landscape. However, the maxima inside are well below the level needed to remove an electron to infinity. This energy difference is associated with the work function of the material.

When the number of atoms is sufficiently large, one can forget about the boundaries and only consider an infinite periodic array of atoms inside the material, as shown in Figure 2.3. This is the basis for electronic structure in solid state physics.

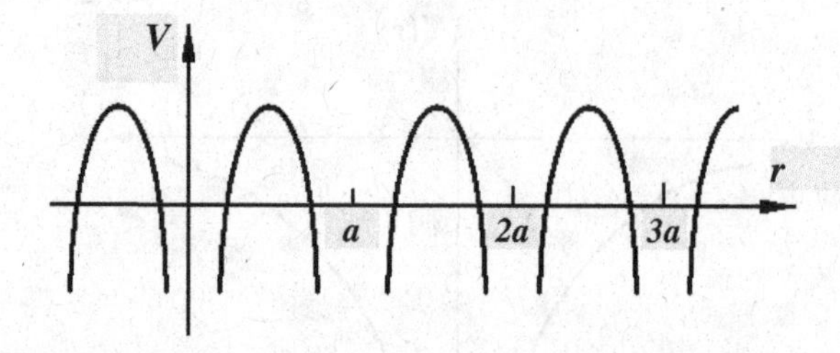

Figure 2.3 Potential energy landscape for an infinite periodic array.

As shown in Figure 2.4, the presence of other atoms broadens the energy levels. The bottom electrons are hardly affected, while the upper electron levels become overlapping energy bands.

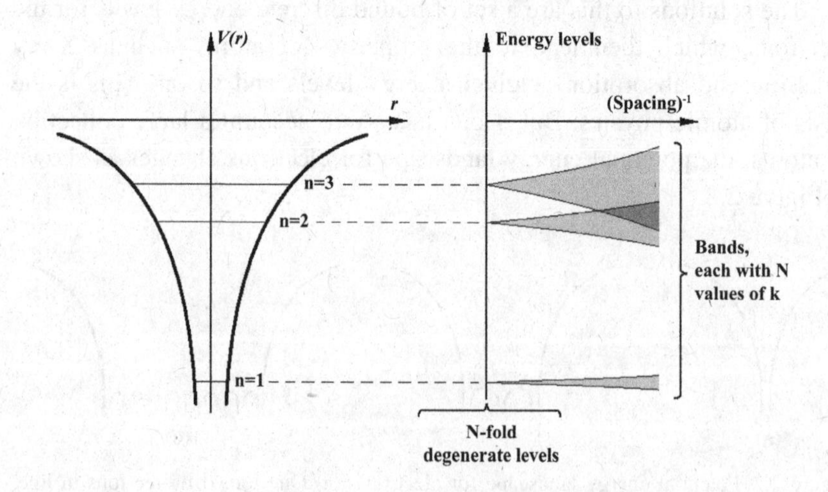

Figure 2.4 Energy bands.

2.2.1.2 *Solids*

The electronic properties of a solid are closely related to its energy bands formed when atoms are brought together to constitute the solid.

(1) *Insulators, semiconductors and conductors*

When a solid is formed, the energy levels of the atoms broaden and form bands with forbidden gaps between them. Electrons can have energy values existing within one of the bands, but cannot have energy values in the gaps between the bands. Usually the lower energy bands at the inner atomic levels are narrow and are all full of electrons, and they do not contribute to the electronic properties of a material. The outer or valence electrons that bond the crystal together occupy what is called a valence band. For an insulating material, there are no delocalized electrons to carry electric current. The conduction band is far above the valence band in energy, as shown in Figure 2.5 (a), so it is not thermally accessible, and remains essentially empty. In other words, the heat content of an insulating material at room temperature (300 K) is not sufficient to raise an appreciable number of electrons from the valence band to the conduction band; therefore the number of electrons in the conduction band is negligible. In other words, for an insulator, the value of the gap energy E_g far exceeds the value of the thermal energy k_BT, where k_B is Boltzmann's constant.

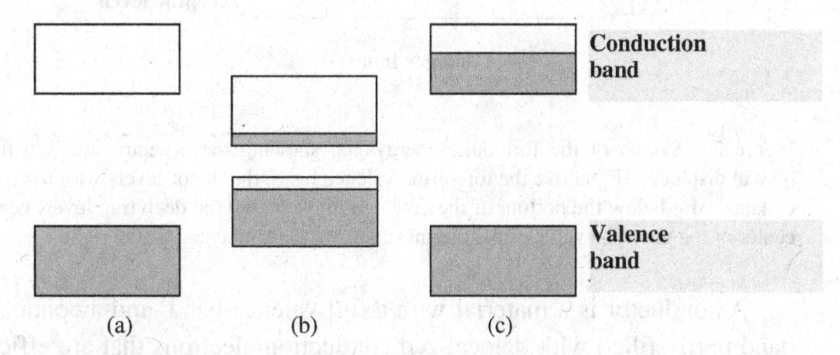

Figure 2.5 Energy bands of (a) an insulator, (b) an intrinsic semiconductor and (c) a conductor. The cross-hatching indicates the presence of electrons in the bands.

As shown in Figure 2.5 (b), for semiconductors, the gap between the valence band and the conduction band is small. As the energy gap E_g is close to the thermal energy k_BT, the heat content of the material at room temperature can thermally excite some electrons from the valence band to the conduction band. As the density of the electrons reaching the conduction band by this thermal excitation process is quite low, the electrical conductivity is very small, so the term "semiconducting" is used. A material of this type is called an intrinsic semiconductor.

Intrinsic semiconductors can be doped with donor atoms that give electrons to the conduction band. The material can also be doped with acceptor atoms that obtain electrons from the valence band and leave behind positive charges, called holes, which can also carry electric current. The energy levels of these donors and acceptors lie in the energy gap, as shown in Figure 2.6. The former produces n-type semiconductors, which have negative-charge carriers, and the latter produces p-type semiconductors, which have positive-charge or hole carriers. The conductivities of these semiconductors are temperature-dependent.

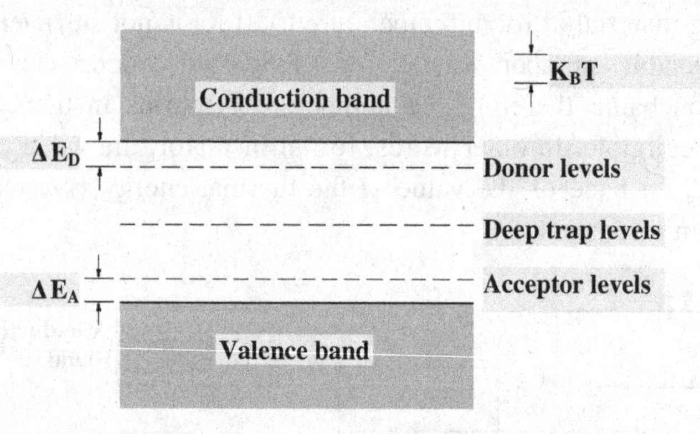

Figure 2.6 Sketch of the forbidden energy gap showing the acceptor levels with the typical distance ΔE_A above the top of the valence band, the donor levels with the typical distance ΔE_D below the bottom of the conduction band, and the deep trap levels near the center of the gap. The value of the thermal energy k_BT is indicated on the right.

A conductor is a material with a full valence band, and a conduction band partly filled with delocalized conduction electrons that are efficient in carrying electric current. The metal ions with positive charge at the lattice sites have given up their electrons to the conduction band and constitute a background of positive charges for the delocalized electrons. Figure 2.5 (c) shows the energy bands for this case.

(2) *Energy bands and gaps of semiconductors*

In actual crystals, the energy bands are much more complicated than what we discussed above. The electrical, optical and other properties of actual semiconductor crystals strongly depend on how the energy of the delocalized electrons involves the wavevector k. Here we consider three-dimensional crystals, and we concentrate on the properties of the III-V and the II-VI semiconducting compounds. These compounds have a cubic structure: their three lattice constants are the same ($a = b = c$).

At the temperature of absolute zero degree (0 K), all the energy bands below the gap are filled with electrons, and all the bands above the gap are empty, so the material is an insulator. At room temperature, as the gap is small, some electrons are thermally excited from the valence band to the conduction band, and these excited electrons gather in the region of the conduction band immediately above its minimum at T_m, a region that is referred to as a "valley". These electrons carry some electric current, so the material is a semiconductor. Gallium arsenide is called a direct-bandgap semiconductor because the top of the valence band and the bottom of the conduction band are both at the same center point (T) in the Brillouin zone. The compounds GaAs, GaSb, InP, InAs and InSb and all the II-VI compounds included in the periodic table have direct gaps. More information about the periodic table can be found in Section 2.4. In some semiconductors such as Si and Ge, the top of the valence band is at a position in the Brillouin zone, which is different from that for the bottom of the conduction band, and they are called indirect-bandgap semiconductors.

2.2.2 *Electron levels for nanostructures*

Considering hundreds of atoms in a single row, it is reasonable to assume that the outermost electrons in conduction bands are almost unbound to their nuclei, and move as if they were in a constant background potential.

For free electrons, the potential is zero. Electrons can also be in a finite square well potential, or an infinite potential well.

2.2.2.1 *Free electron*

For a free electron, the potential *V(x)* is zero everywhere. For the one-dimensional case, the Schrödinger equation reduces to:

$$\frac{h^2}{2m}\cdot\frac{\partial^2\Psi}{\partial x^2} = E\Psi \text{ for all } x \tag{2.8}$$

In the k domain, the equation becomes:

$$h^2\, k^2\Psi = E\,\Psi \tag{2.9}$$

or

$$E = p^2/2m \tag{2.10}$$

This is identical to the classical free electron energy equation. Therefore, in the free-electron case, the quantum mechanics solution is identical to the classical mechanics solution.

2.2.2.2 *Electron in a finite potential well*

The potential well is assumed to exist from –a to +a. In other words, the potential *V* is zero in the range from –a to +a, and the potential *V* is finite from $-\infty$ to –a and from +a to $+\infty$. The Schrödinger equations for the three regions can be written as:

$$\frac{h^2}{2m}\frac{\partial^2\Psi(x)}{\partial x^2} + V(x)\,\Psi(x) = E\,\Psi(x) \qquad -\infty < x < -a \tag{2.11}$$

$$\frac{h^2}{2m}\frac{\partial^2\Psi(x)}{\partial x^2} = E\,\Psi(x) \qquad -a < x < +a \tag{2.12}$$

$$\frac{h^2}{2m}\frac{\partial^2\Psi(x)}{\partial x^2} + V(x)\,\Psi(x) = E\,\Psi(x) \qquad a < x < +\infty \tag{2.13}$$

By applying boundary conditions and making the following substitutions:

$$\alpha^2 = \frac{2m}{h^2}(V - E) \text{ and } \beta^2 = \frac{2m}{h^2}E \tag{2.14}$$

the Schrödinger equations in the three regions reduce to:

$$\frac{\partial^2\Psi(x)}{\partial x^2} - \alpha^2\Psi(x) = 0 \qquad -\infty < x < -a \tag{2.15}$$

$$\frac{\partial^2\Psi(x)}{\partial x^2}+\beta^2\Psi(x)=0 \quad -a < x < +a \tag{2.16}$$

$$\frac{\partial^2\Psi(x)}{\partial x^2}-\alpha^2\Psi(x)=0 \quad a<x<+\infty \tag{2.17}$$

2.2.2.3 *Electron in an infinite potential well*

Another typical case is the bound electron case where the electron is contained within a potential well of infinite depth. For this case, the solutions of the one-dimensional Schrödinger equation become:

$$\Psi(x)=0 \quad -\infty<x<-a \tag{2.18}$$

$$\Psi(x)=B_1\sin(\beta x)+B_2\cos(\beta x) \quad -a < x < +a \tag{2.19}$$

$$\Psi(x)=0 \quad +a<x<+\infty \tag{2.20}$$

By applying the boundary conditions at $-a$ and $+a$, we get the following eigen equation:

$$\sin(\beta\alpha)=0 \tag{2.21}$$

By further substitutions, we obtain the following result,

$$E=\frac{n^2\pi^2h^2}{2ma^2} \tag{2.22}$$

where m is the mass of the electron, and a is the width of the potential well.

2.2.2.4 *Schrödinger equations*

Here we discuss the Schrödinger equations for three special cases: one-electron atom, multi-electron atom and molecule.

(1) *One-electron atom*

Schrödinger built the relation between classical waves and de Broglie's particle waves. By substituting the de Broglie wavelength for λ in the classical wave equation to adapt the situation to the particle waves, we obtain the following Schrödinger equation:

$$\nabla^2\psi=-(2\pi/\{h/[2m(E-V)]^{1/2}\})^2\psi=-[8\pi^2m/h^2(E-V)]\psi \tag{2.23}$$

This equation can be rearranged as

$$H\psi=E\psi \tag{2.24}$$

For a particle with charge q_1 in the field of another particle with charge q_2, its electrostatic potential energy, V, is given by:

$$V = q_1 q_2 / r \qquad (2.25)$$

where r is the distance between these two charged particles. The potential energy V between an electron ($q_1 = -e$) and a nucleus ($q_2 = +Ze$) can be written as:

$$V = -Ze^2 / r \qquad (2.26)$$

By substituting Eq. (2.26) for V in the wave equation, we have:

$$[(-h^2 / 8\pi^2 m_e)\nabla^2 - Ze^2 / r]\psi = E\psi \qquad (2.27)$$

Each Ψ and its corresponding energy correspond to an electron bound to a nucleus with charge $+Ze$. The function Ψ is usually called one-electron orbital, and is also known as hydrogen like atomic orbital. Such orbital can be analytically obtained by solving the Schrödinger equation in closed form.

(2) *Multi-electron atom*

For an atom with k electrons, by considering the electron states, the interactions between the electron and the nucleus, and the electron-to-electron interactions, its Hamiltonian can be expressed as:

$$H = \left[\left(\frac{-h^2}{8\pi^2 m_e}\right)\sum_{i=1}^{k}\nabla^2 - \sum_{i=1}^{k}\frac{z}{r_1}\right] + \sum_{i=1}^{k-1}\sum_{j=1+i}^{k-1}\frac{1}{r_{ij}} \qquad (2.28)$$

The Schrödinger equation for a multi-electron atom is usually solved in a numerical way.

(3) *Molecule*

According to Born-Oppenheimer approximation, in a molecule with N atoms and k electrons, the nuclei can assumed to be stationary. By considering the energies of the electrons, the electron-to-nucleus interactions, the electron-to-electron interactions and the nucleus-to-nucleus interactions, the Hamiltonian for a molecule with N atoms and k electrons is given by:

$$H = \left[\left(\frac{-h^2}{8\pi^2 m_e}\right)\sum_{i=1}^{k}\nabla^2 - \sum_{j=1}^{N}\sum_{i=1}^{k}\frac{z_j}{r_{ji}}\right] + \sum_{i=1}^{k-1}\sum_{l=1+i}^{k-1}\frac{1}{r_{il}} + \sum_{j=1}^{k-1}\sum_{j=1+i}^{k-1} Z_j Z_m / R_{jm} \qquad (2.29)$$

2.2.3 *Density of states and confinement*

To achieve confinement, the nanostructure length scale a should be less or close to the characteristic length scale L of the electron or hole behavior in a normal bulk matter. According to the relationship between a and L, one can further distinguish between strong confinement and weak confinement

$$a < L \qquad \text{strong confinement} \qquad (2.30)$$
$$a > L \qquad \text{weak confinement} \qquad (2.31)$$
$$a >> L \qquad \text{no confinement} \qquad (2.32)$$

Confinement in a three-dimensional nanoparticle means that the electrons essentially have no freedom to move. However in a thin film, electrons are free to move in two dimensions, but are confined in one dimension. The delocalized dimensions and confined dimensions of quantum dots, quantum wires, quantum wells and bulks are listed in Table 2.1.

Table 2.1 Dimensions of quantum structures.

Type	Delocalized dimensions	Confined dimensions
Quantum dot	0	3
Quantum wire	1	2
Quantum well	2	1
Bulk	3	0

The density of states D is an important quantity for understanding various spectroscopic and transport properties of nanomaterials. It is the density of states that makes the electronic properties of nanostructures different from bulk matters. The density of states D is usually defined as the number of states per unit energy range:

$$D(E) = dN(E)/dE \qquad (2.33)$$

where $N(E)$ is the possible number of electrons. Table 2.2 lists the density of states and number of electrons for quantum wire, quantum well and bulk.

The density of states affects a lot of electronic properties of a matter. In the 0D, 1D and 2D confined systems, the electron levels are discrete in the confined dimension. The differences in the density of states result

Table 2.2 Dimensions of states and number of electrons for quantum structures.

Type	Delocalized dimensions	D(E)	N(E)
Quantum wire	1	$\alpha\ E^{-1/2}$	$\alpha\ E^{-1/2}$
Quantum well	2	Independent of E	α E
Bulk	3	$\alpha\ E^{1/2}$	$\alpha\ E^{3/2}$

in great differences in the electronic properties of 0, 1 and 2D confined electron systems as compared to bulks [52].

2.2.3.1 *Zero-dimensional solid*

Neglecting the periodic potential existing in solids, one can imagine a zero-dimensional (0D) solid in which electrons are confined in a three-dimensional potential box with extremely small (<100 nm) length, breadth and height. This particle has discrete energy levels as discussed above with density of states given by:

$$D(E) = \frac{dN}{dE} = \sum_{\varepsilon_i} \delta\left(E-\varepsilon_i\right) \tag{2.34}$$

where ε_i are discrete energy levels and δ is Dirac function. Figure 2.7 illustrates the density of states as a function of energy.

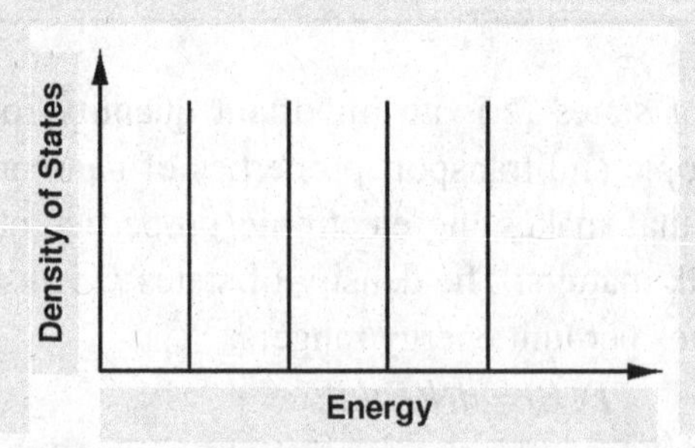

Figure 2.7 Density of states for a particle in a zero-dimensional solid [52].

2.2.3.2 *One-dimensional quantum wire*

A quantum wire is also called a one-dimensional (1D) solid. For a quantum wire, the potential in two dimensions is infinitely large, while in the third dimension, the potential is zero. The density of states of a quantum wire can be expressed as:

$$D(E) = \frac{dN}{dE} = \sum_{\varepsilon_i < E} \delta(E - \varepsilon_i) \qquad (2.35)$$

where ε_i represents a discrete energy level. Figure 2.8 graphically shows the nature of density of states for a quantum wire.

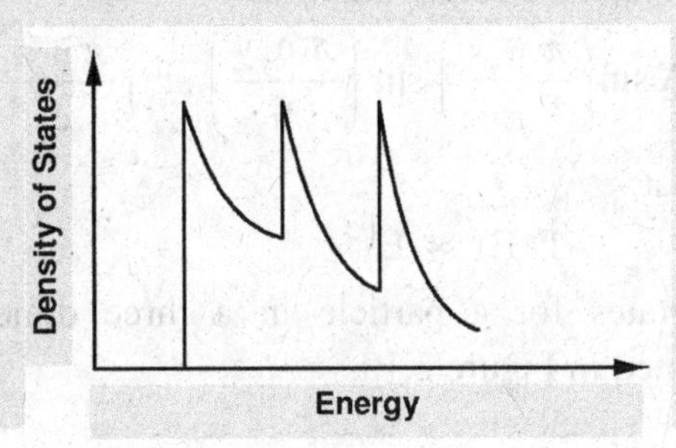

Figure 2.8 Density of states for a particle in a one-dimensional quantum wire [52].

2.2.3.3 *Two-dimensional thin film*

A thin film can be taken as a two-dimensional solid. The density of states *D(E)* of a thin film is given by:

$$D(E) = \frac{dN}{dE} = \sum_{\varepsilon_i < E} 1 \qquad (2.36)$$

As shown in Figure 2.9, the density of states in a thin film is a staircase.

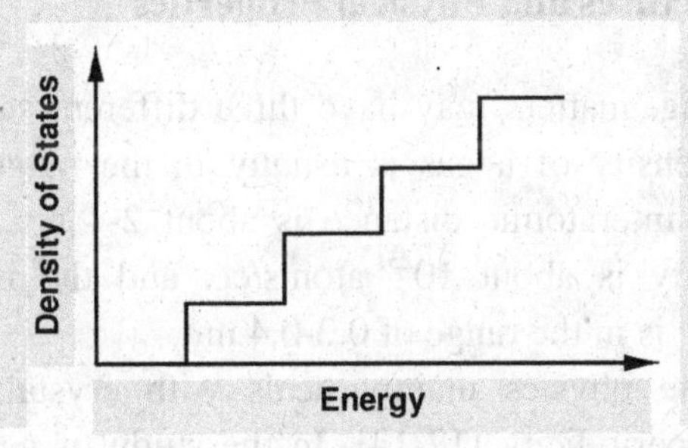

Figure 2.9 Density of states for a 2D potential box [52].

2.2.3.4 *Particle in a three-dimensional box*

Here we discuss a particle in a box with length a, width b and height c. The potential V is zero inside the box, while infinite outside the box. The energy states of such a particle are given by:

$$E_{n_x,n_y,n_z} = \frac{h^2}{2m}\left(n_x^2 + n_y^2 + n_z^2\right) \tag{2.37}$$

The wave function is in the following form:

$$\Psi_n(x,y,z) = A \sin\left(\frac{\pi n_x x}{a}\right) \sin\left(\frac{\pi n_y y}{b}\right) \sin\left(\frac{\pi n_z z}{c}\right) \tag{2.38}$$

It can be shown that:

$$D(E) \propto E^{1/2} \tag{2.39}$$

The density of states for a particle in a three dimensional box is graphically illustrated in Figure 2.10.

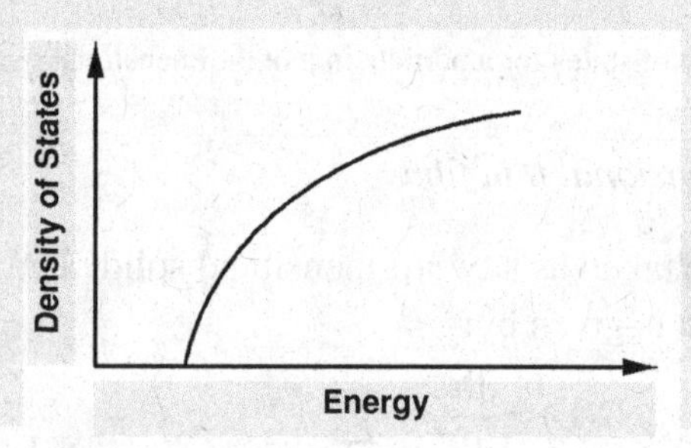

Figure 2.10 Density of states for a particle in a 3D potential box [52].

2.3 Crystal Structures and Physical Properties

Generally speaking, matters may have three different states: gas, liquid and solid. The density of a gas is usually in the range of about 10^{19} atoms/cc and the interatomic distance is about 2-4 nm. For solids and liquids, the density is about 10^{23} atoms/cc, and the inter-atomic and molecular distance is in the range of 0.2-0.4 nm.

The solid-state physics mainly deals with crystals whose atoms exhibit excellent periodicity [12-14]. In the study of nanocrystals, it is important to know its characteristic features, such as electrical, magnetic, optical or mechanical properties, which can be exploited for engineering applications [15-17].

Each crystal exhibits a unique arrangement of atoms, called crystal structure. Most of the characteristic properties of a crystal are related to its crystal structure. In discussing crystal structures, three concepts are often used: lattice, unit cell, and basis [18-24], and their relationships are schematically shown in Figure 2.11.

Generally speaking, a lattice is a regular arrangement of points in space. It is not an arrangement of atoms, but an arrangement of points which form the framework onto which the atoms are hung [25-30]. A lattice is often defined as points repeated in one dimension, two dimensions or three dimensions, making it a 1D, 2D or 3D lattice. A lattice can be taken as unit cells stacked together. A unit cell is defined as a set of points arranged in a particular way, and the periodic repetition of unit cells in one dimension, two dimensions, or three dimensions forms a 1D, 2D, or 3D lattice. The spacing between the points in each direction is called the lattice parameter in this direction. A unit cell can be described by its lattice parameters and the angles between them. It should be noted that different cell units may form the same lattice, and among the various unit cells forming a lattice, the smallest unit cell is called the primitive cell.

In order to describe a crystal structure completely, one should specify the underlying lattice and the basis associated with each point in the lattice. A basis is an atom or a group of atoms associated with each lattice point, and it is usually described by the types and numbers of the atoms in the basis and their positions. A complete description of a crystal structure includes both the lattice and the basis.

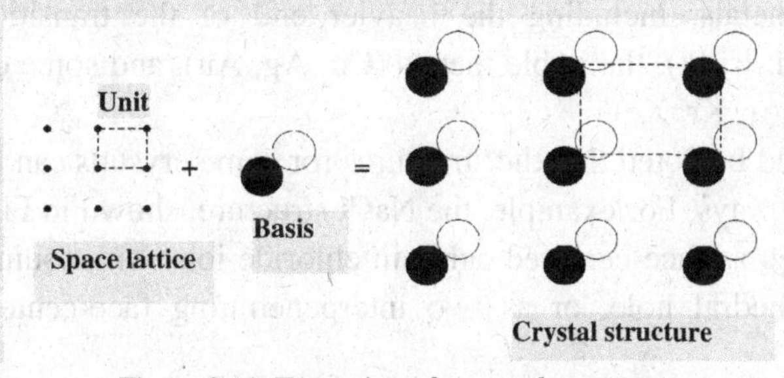

Figure 2.11 Formation of a crystal structure.

2.3.1 *Cubic crystal lattice structures*

In a cubic crystal system, the unit cell is in the cube shape. For most metallic crystals, the unit cell with cubic shape is the simplest and the most common. As shown in Figure 2.12, simple cubic (SC), body-centered cubic (BCC) and face-centered cubic (FCC), are three typical kinds of arrangements in the cubic crystal system.

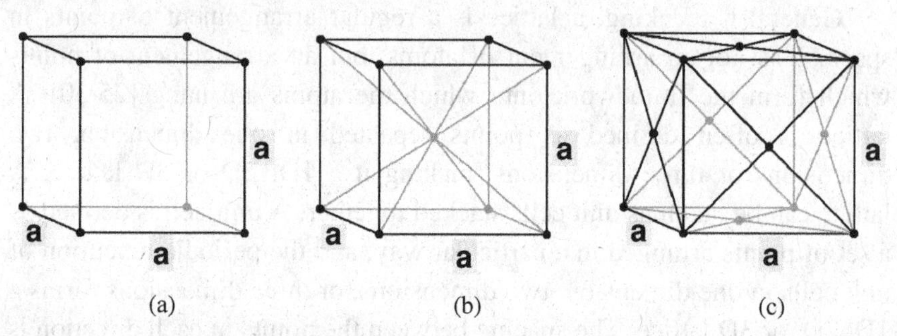

Figure 2.12 Three typical kinds of cubic crystal system. (a) Simple cubic (SC), (b) body-centered cubic (BCC), (c) face-centered cubic (FCC).

In a simple cubic (SC) system, there is one lattice point on each corner of the cube. In a body-centered cubic (BCC) system, in addition to the eight corner points, there is one lattice point in the center of the unit cell. The BCC structure is common for the elements at the lighter end of all the rows of the periodic table, from the alkali metals (Li, Na, K, Rb, Cs) through to the start of the transition metals (V, Cr, Fe, Nb, Mo, Ba, Ta, W). In a face-centered cubic (FCC) lattice, in addition to the eight corner points, there is one lattice point on the center of each face, giving a total of 14 lattice points. The FCC structure is adopted by about 20 types of metals, including the heavier end of the transition metals (Ni, Rh, Pd, Ir, Pt), the noble metals (Cu, Ag, Au), and some of the rare gases (Ne, Ar, Kr, Xe).

It should be noted that the structures for some crystals can be viewed in different ways. For example, the NaCl structure, shown in Figure 2.13, can be taken as face-centered cubic in chloride ions with sodium ions in every octahedral hole, or as two interpenetrating face-centered cubic structures.

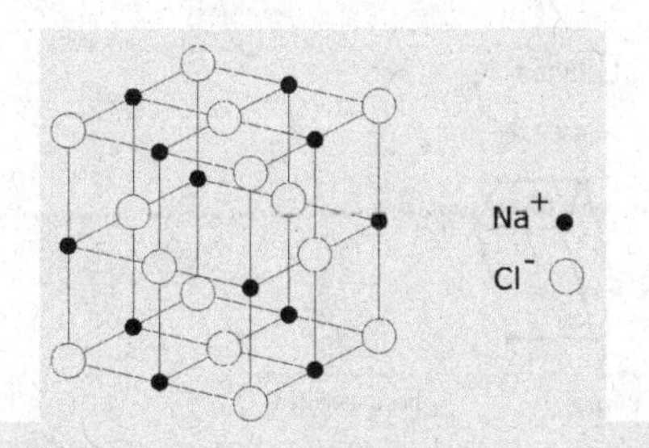

Figure 2.13 Crystal structures of NaCl [52].

2.3.2 *Symmetry in crystals*

A crystal has its inherent symmetry, which means that, the crystal remains unchanged under certain operations. For example, by rotating a crystal with two-fold rotational symmetry 180° about its symmetrical axis, an atomic configuration which is the same as the original configuration can be obtained. In addition to rotational symmetries, there are other types of symmetries, such as mirror symmetries, translational symmetries and compound symmetries which are a combination of translation and rotation/mirror symmetries. Figure 2.14 shows a full classification of crystals, from which all the inherent symmetries of crystals can be identified.

Twenty of the 32 crystal classes are so-called piezoelectric, and crystals belonging to one of these classes (point groups) display piezoelectricity. All the 20 piezoelectric classes lack a center of symmetry. Any material develops a dielectric polarization when an electric field is applied, but a substance which has such a natural charge separation even in the absence of a field is called a polar material. Whether or not a material is polar is determined solely by its crystal structure. Only 10 of the 32 point groups are polar. All the polar crystals are pyroelectric, so the 10 polar crystal classes are sometimes referred to as pyroelectric classes.

Real crystals feature defects or irregularities in the ideal arrangements described above and it is these defects that critically determine many of the electrical and mechanical properties of real materials. In particular, dislocations in the crystal lattice allow shear at much lower stress than that needed for a perfect crystal structure.

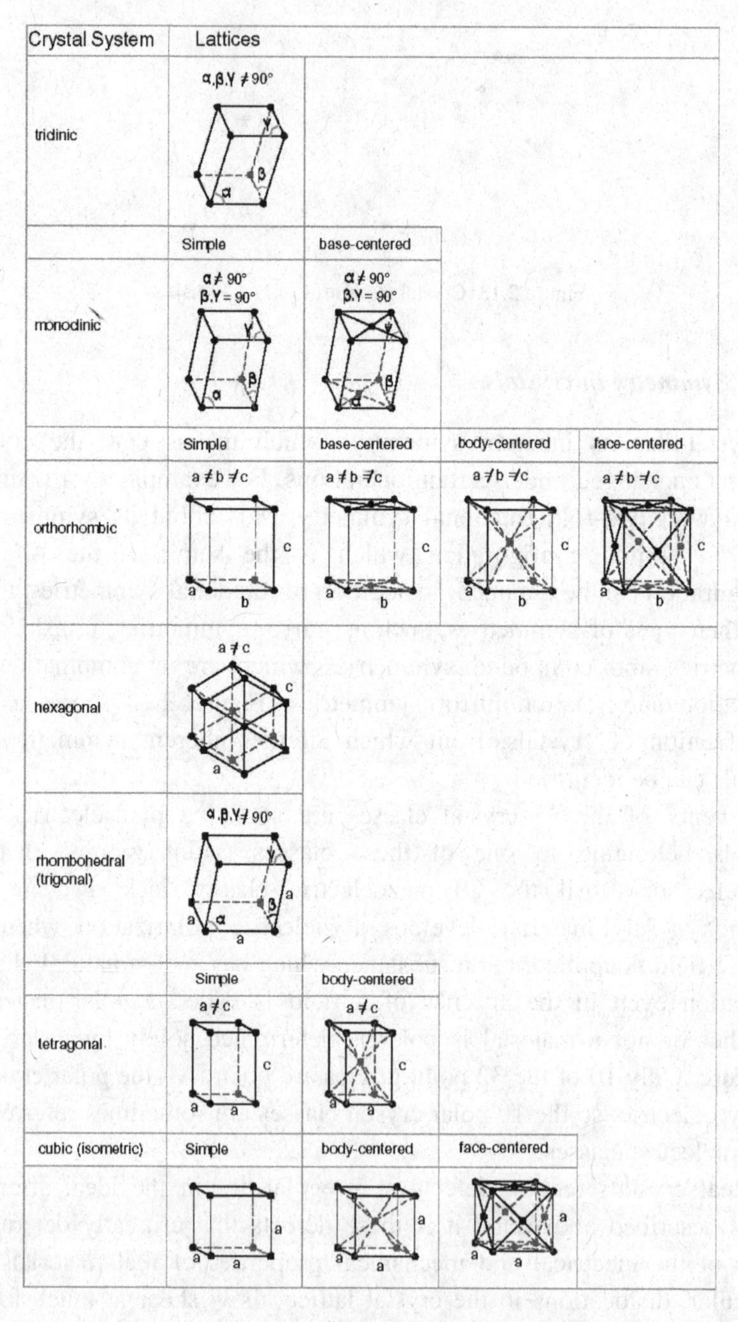

Figure 2.14 The crystal system and bravais lattice.

2.3.3 *Bravais lattices and point/space groups*

Crystal systems are crystal structures organized according to the axial systems used to describe their lattices. There are seven crystal systems, each of which is described by a set of three axes in a particular geometrical arrangement. The cubic (or isometric) system is the simplest and most symmetric system. It has cubic symmetry: the three axes have equal length, and are mutually perpendicular. In order of decreasing symmetry, other six systems are hexagonal, tetragonal, rhombohedral, orthorhombic, monoclinic and triclinic.

By combining crystal systems with possible lattice centering, we get the Bravais lattices. The Bravais lattices are sometimes referred to as space lattices. They describe the geometric arrangement of the lattice points, and thereby the translational symmetries of crystals. As shown in Figure 2.14, in a three-dimensional space, there are 14 unique Bravais lattices, and all crystals recognized so far fit in one of these arrangements.

2.3.4 *Quasicrystals*

Quasicrystals, a peculiar type of solid, were first discovered by Dan Shechtman in 1982. In a quasicrystal, atoms are arranged in a seemingly regular, yet nonrepeating way. Most of the quasicrystals are aluminum alloys, for example, Al-Ni-Co, Al-Pd-Mn and Al-Cu-Fe, while quasi-crystals with other compositions are also available, such as Ti-Zr-Ni and Zn-Mg-Ho. Though the local arrangements of atoms a quasicrystal are fixed and in a regular pattern, such arrangements are not periodic throughout the entire material. Each cell in a quasicrystal may have a configuration which is different from of those of its surrounding cells.

2.3.5 *Liquid crystals*

Liquid crystals (LCs) exhibit properties between those of a solid crystal and those of a conventional liquid. Like a liquid, a liquid crystal may flow, but its molecules are arranged and oriented in a crystal-like way. Different types of LC phases may have different optical properties.

Under a polarized light, a liquid crystal will appear to have a distinct texture. The "patch" in the texture corresponds to the LC molecules that are oriented in a different direction. However, within a domain, the molecules are well ordered.

Just as water has different phases, a liquid crystal material may exist in the solid or gas phase. There are two types of liquid crystals: thermotropic LCs and lyotropic LCs. Thermotropic LCs exhibit a phase transition when the temperature changes, while lyotropic LCs exhibit phase transitions as a function of concentration. Molecules exhibiting liquid crystal phases are usually called mesogens. For a molecule in an LC phase, it should be rigid and anisotropic. Generally speaking, there are two types of mesogens: calamitic mesogens and discotic mesogens. Calamitic mesogens have a rigid-rod shape, and they orient along their long axis. Discotic mesogens are disk-like, and they orient along their short axis.

In living systems, there are numerous lyotropic liquid-crystalline nanostructures. For example, cell membranes and biological membranes are a type of liquid crystal. Membranes are elastic and fluid, and their constituent rod-like molecules, such as phospholipids, are perpendicular to the membrane surfaces. The constituent molecules can easily flow in-plane, but they do not leave the membrane. However, with some difficulty, the constituent molecules can flip from one side of the membrane to the other side. Furthermore, many important proteins, such as receptors, can freely "floating" inside, or partly outside, the membranes.

There are many other biological structures exhibiting LC properties. For example, the protein solution extruded by a spider to generate silk is a kind of liquid crystal. The precise ordering of molecules in silk is important to its excellent strength. DNA and many polypeptides also exhibit LC properties. Biological mesogens usually exhibit chiral behaviors, and chirality often plays an important role in the liquid crystals formed by biological mesogens.

Liquid crystals are widely used in liquid crystal displays (LCDs), which rely on the optical properties of certain liquid crystalline molecules in the presence or absence of an electric field. In a typical device, a liquid crystal layer sits between two polarizers that are crossed, being oriented at 90° to one another. The liquid crystal is chosen so that its relaxed phase is a twisted one. This twisted phase reorients light that has passed through the first polarizer, allowing it to be transmitted through the second polarizer and reflected back to the observer. The device thus appears clear. When an electric field is applied to the LC layer, all the mesogens align, and are no longer twisting. In this aligned state, the mesogens do not reorient light, so the light polarized at the first polarizer is absorbed at the second polarizer, and the entire device appears dark. In this way, the electric field can be used to make a pixel switch between clear or dark on command. Color LCD systems use the same technique, with color filters used to generate red, green and blue pixels. Similar principles can be used to make other liquid crystal based optical devices. Some liquid crystal materials change colors when stretched or stressed. Such liquid crystal sheets are often used in industry to look for hot spots, map heat flow, measure stress distribution patterns and so on. Liquid crystal in fluid form can be used to detect electrically generated hot spots for failure analysis in the semiconductor industry.

2.4 Physical Aspects of Nanochemistry

In the research of nanoscience and nanotechnology, it is important to consider the physical aspects of nanochemistry, especially the surface energy of nanostructured materials.

2.4.1 *Nomenclature and periodic table of elements*

Elements are the fundamental substances from which all matters are composed. Currently, there are 109 kinds of officially named elements. The tentative 110th element has been synthesized, and elements with higher atomic numbers could be synthesized in the future.

Elements can be organized according to a periodic table based on their atomic numbers, where elements share some of their physical and chemical properties. In the mid-19th century, researchers realized that elements could be grouped according to their chemical behaviors. For example, some elements have similar characteristics: helium (He), neon

(Ne) and argon (Ar) are nonreactive gases, while lithium (Li), sodium (Na) and potassium (K) are very reactive metals. Along with the increase of the atomic number, elements exhibit periodic changes in their chemical and physical properties. This kind of investigation resulted in the periodic table of elements, as shown in Figure 2.15.

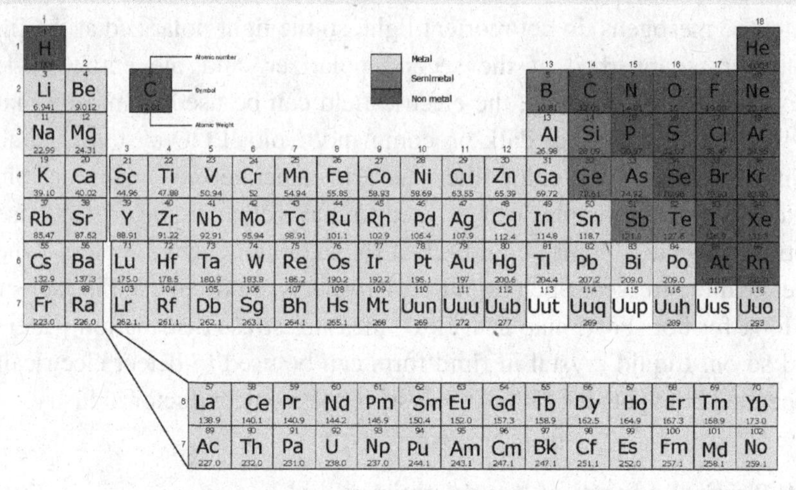

Figure 2.15 Periodic table of elements by atomic number.

In the periodic table, periods are shown as horizontal rows and groups are shown in vertical columns. The elements in the same period have the same number of orbitals. Periods are characterized by the number of energy levels (shells) of the electrons surrounding the nucleus. The elements in first period have one shell and have at most two electrons, while the elements in second period have two shells and have at most 10 electrons, and so on. The elements in the same group have the same number of electrons in the tilling orbital, and they have similar properties.

2.4.2 *Surface energy of nanostructured materials*

Nanostructured materials possess a large fraction of surface atoms per unit volume [31-34]. The ratio of surface atoms to the interior atoms increases dramatically if one successively divides a macroscopic object into smaller parts. The total surface energy increases with the overall surface area, which is in turn strongly dependent on the dimension of the material. The surface area and the surface energy are negligible when cubes are large, but become significant for very small particles. When the size of a particle changes from centimeter to nanometer, the surface area and the surface energy increase seven orders of magnitude.

Due to their extremely large surface areas, nanostructured materials possess huge surface energies and, thus, are thermodynamically unstable or metastable. One of the great challenges in the fabrication of nanomaterials is to overcome the surface energy, and to prevent the nanostructures or nanomaterials from growth in size, driven by the reduction of overall surface energy. In order to produce and stabilize nanostructures and nanomaterials, it is essential to have a good understanding of surface energy and surface physical chemistry of solid surfaces [6, 35].

Atoms or molecules on a solid surface possess fewer nearest neighbors, and thus have dangling bonds exposed to the surface. Because of the dangling bonds on the surface, surface atoms or molecules are under an inwardly directed force and the bond distance between the surface atoms or molecules and the sub-surface atoms or molecules, is smaller than that between interior atoms or molecules. When solid particles are very small, such a decrease in bond length between the surface atoms and interior atoms becomes significant and the lattice constants of the entire solid particles show an appreciable reduction. The extra energy possessed by the surface atoms is described as surface energy, surface free energy or surface tension. By definition, surface energy γ is the energy required to create a unit area of new surface:

$$\gamma = (\partial G/\partial A)\, n_i, T, P \qquad (2.40)$$

where A is the surface area.

As illustrated in Figure 2.16, when a solid material is divided into two pieces, two new surfaces are created. On the newly created surfaces, each atom is located in an asymmetric environment and will move towards the interior due to the breaking of bonds at the surface. An extra force is required to pull the surface atoms back to its original position. Such a surface is called a singular surface.

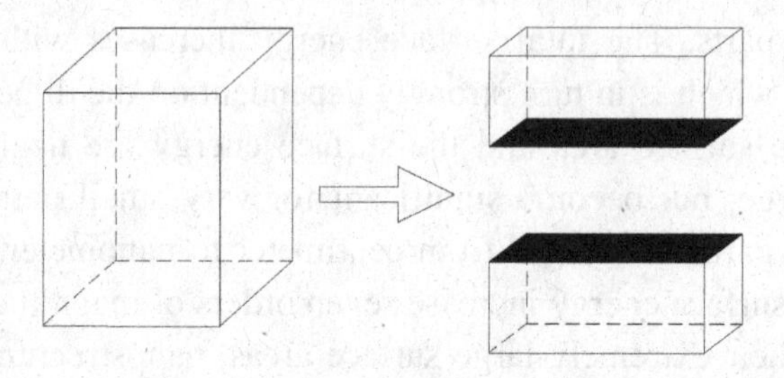

Figure 2.16 Two new surfaces created by breaking a solid into two pieces.

For each atom on such a singular surface, the energy required to get it back to its original position will be equal to the number of broken bonds N_b multiplied by half of the bond strength ε. Therefore, the surface energy is given by:

$$\gamma = \frac{1}{2} N_b \varepsilon \rho_a \tag{2.41}$$

where ρ_a is the surface atomic density, and N_b is the number of atoms per unit area on the new surface. This basic model ignores the interactions owing to higher order neighbors, assumes that the ε value is the same for surface and bulk atoms, and does not include entropic or pressure-volume contributions. Equation (2.41) gives a rough estimation of the true surface energy of a solid surface, and is only applicable to solids with rigid structure where no surface relaxation occurs. When there is an appreciable surface relaxation, such as the surface atoms moving inwardly, or there is a surface restructuring, surface energy is lower than the value given by Eq. (2.41). In spite of the overly simplified assumptions used, this equation does provide some general guidance [6].

According to thermodynamics, a material or system is stable only when it is in a state with the lowest Gibbs free energy. Therefore, there is a strong tendency for a solid or a liquid to minimize its total surface energy. There are a variety of mechanisms to reduce the overall surface energy, and these mechanisms can be grouped into atomic or surface level, individual structures and the overall system [36-38]. For a given surface with a fixed surface area, the surface energy can be reduced by following approaches [39, 40]:

- Surface relaxation. In this approach, the surface atoms or ions shift inwardly. Due to the rigid structures in solids, surface relaxation occurs more readily in a liquid surface than in a solid surface.
- Surface restructuring through combining surface dangling bonds into strained new chemical bonds.
- Surface adsorption through chemical or physical adsorption of terminal chemical species onto the surface by forming chemical bonds or weak attraction forces such as electrostatic or van der Waals forces.
- Composition segregation or impurity enrichment on the surface through solid-state diffusion.

2.5 Bionanoscience and Bionanotechnology

2.5.1 *Basic concepts*

Everything in this universe, from giant stars to our bodies, works on a molecular scale. Our hearts and lungs are big objects but all the processes that make them work perform at the molecular level. Due to their attractive attributes, biological molecules and systems are highly suitable for nanotechnology applications. For instance, proteins precisely fold into three-dimensional shapes, and nucleic acids assemble following well-defined rules. Antibodies recognize and bind their ligands in a highly specific way, and biological assemblies, such as molecular motors, can accurately perform extremely complicated operations.

The controlled design of biological molecules on the nanometer scale to achieve desired structures or functionalities is an active research topic in bionanoscience. In addition to the actual control of biological molecules, bionanoscience also involves, for instance, human implant coatings, modification of stent inner surfaces to change blood adhesion, and lab-on-chips for analysis of biological substances. Furthermore, bionanoscience also studies the interactions between living organisms and nanostructures, and biomimetics, which deals with materials that somehow mimic natural materials, is an important branch of bionanoscience.

The design principles that biology uses most effectively are those that exploit at the nanoscale features in an environment of liquid water. These include the highly effective uses of self-assembly using hydrophobic interaction. For industrial applications, the area of green chemistry is gaining significance for the environmentally friendly processing technologies primarily for making complex, nanostructured materials. These applications use water as solvent and work at lower operating temperatures. The use of templating strategies and precursor routes enhances the scope of making final products that are themselves insoluble in water. And this becomes important in the area of nanobiotechnology for any final product to retain its structural integrity in the ambient environment.

Although a lot of research has been conducted on understanding the concepts of viruses, DNA, proteins, etc., with the nanotech interface, the research is now more focused on nanoscale product realizations. The brain of any living being is like a high-performance information-processing system. Not only their facilities, the underlying architectures provide helpful features for the development of nanoelectronics. There exists a strong correlation between some of the basic biological concepts and how these can be integrated with the quantum electrical and mechanical concepts which form the foundations of nanotechnology [41]. Great progresses have been made in this area. One classic example is the biological network where, integration of biological neurons and MOS circuits are carried out on the same substrate.

The US National Institute of Health (NIH) has identified three broad areas of bionanotechnology research:

- Use nanotechnology tools and concepts to study biology;
- Develop biological molecules to function differently from those found in nature;
- Manipulate biological systems by methods more precise than molecular biological, synthetic chemical and biochemical approaches that have been used for years in the biology research community.

2.5.2 *Lipids*

Lipids are one of the most versatile building blocks in biology in constructing three-dimensional structures. They are molecules that possess a hydrophobic hydrocarbon tail and a hydrophilic polar head group. In the language of chemistry, the self-organization of molecules leads to supramolecular systems and is responsible for their functions. Thermotropic and lyotropic liquid crystals are such functional units, formed by self-organization. As highly oriented systems, they exhibit new and improved properties. The lyotropic liquid crystals as known for long are prerequisite for the development of life and the ability of cells to function. All lipids are hydrophobic and this group of molecules includes fats and oils, waxes, phospholipids, steroids and some other related compounds.

Lipids have following three major functions that are of great physiological importance for humans:

- They are the structural components for biological membranes;
- They reserve energy, mainly in the form of triacylglycerols;
- Both lipids and lipid derivatives serve as vitamins and hormones.

Lipids, as well as the synthetic compounds whose hydrophilic head-groups and hydrophobic tails are made up of groups that may never occur in nature, are also known as amphiphilic compounds or amphiphiles, which remain active in both oil and water. It is possible to extract naturally occurring lipids or to synthesize lipids using standard organic chemistry techniques. Literally thousands of different amphiphiles have been extracted, synthesized and investigated. This is important for the precise control required for building nanoarchitectures.

2.5.3 *Proteins*

Proteins are a group of polymers that can be used to switch something on or off, move something, sense something, taste, smell, produce energy, convert sunlight to sugar, nitrogen to ammonia fertilizer, fight diseases and so on.

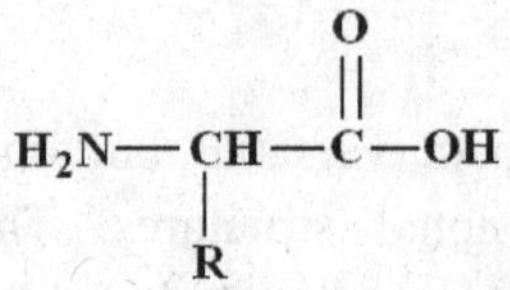

$$H_2N\text{—}CH(R)\text{—}C(=O)\text{—}OH$$

Figure 2.17 Arrangement of individual amino acids. In the figure, "R" represents the side chain functional group. R = hydrophobic, hydrophilic, positively/negatively charged.

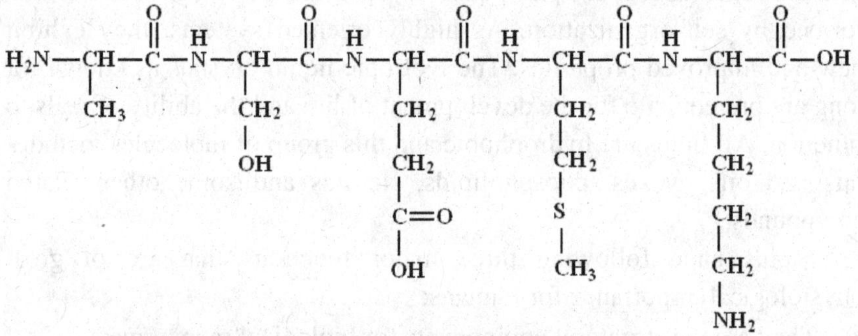

$$H_2N\text{—}CH(CH_3)\text{—}C(=O)\text{—}N(H)\text{—}CH(CH_2OH)\text{—}C(=O)\text{—}N(H)\text{—}CH(CH_2CH_2C(=O)OH)\text{—}C(=O)\text{—}N(H)\text{—}CH(CH_2CH_2SCH_3)\text{—}C(=O)\text{—}N(H)\text{—}CH(CH_2CH_2CH_2CH_2NH_2)\text{—}C(=O)\text{—}OH$$

Figure 2.18 Polypeptide formed by linkage of amino acids.

Proteins are made up of amino acids and there are 20 types of common amino acids. As shown in Figure 2.17, different amino acids basically only differ by the type of side chain functional group. This side chain can be hydrophobic, hydrophilic, positively or negatively charged. As shown in Figure 2.18, biological systems can synthesize linear polymers of these amino acids through amide bonds.

Proteins, which make the basic building blocks of most cells and thus living organisms, are quite complex objects. Proteins can come in a wide variety of sizes and shapes, and several examples are shown in Figure 2.19.

The amazing thing is that out of all the possible structures that the polymer could adapt during the synthesis procedure, the amino acid polymer folds into one specific stable three-dimensional structure [42, 52, 53], as shown in Figure 2.20. This is no simple trick, and so far even using supercomputers we have not been able to figure out how it is done. In other words, given a sequence of amino acids, we cannot predict what structure it will adapt or how to force it to adopt a specific structure.

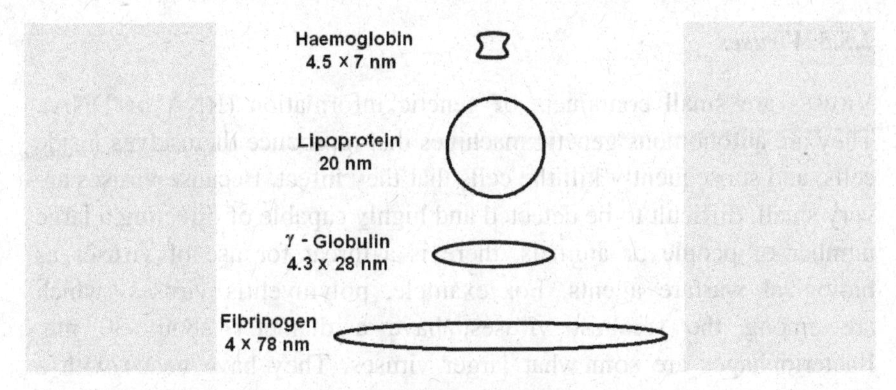

Figure 2.19 Shapes and sizes of several types of proteins. Source: C. P. Poole, Jr. and F. J. Owens, Introduction to Nanotechnology, John Wiley and Sons, New Jersey, © 2003. Reprinted with permission of John Wiley and Sons, Inc.

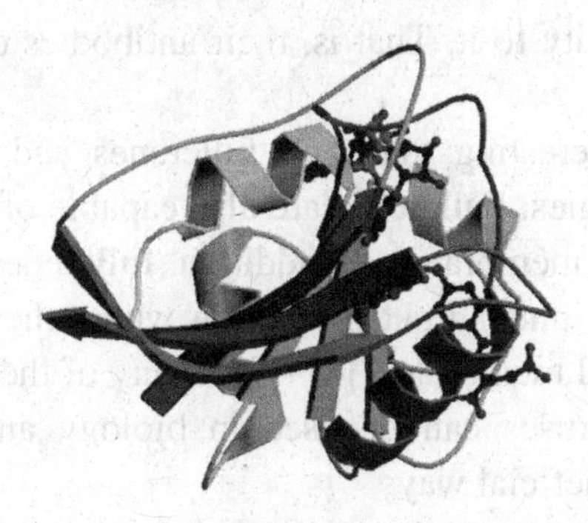

Figure 2.20 Tertiary structure of protein is formed by folding a polypeptide into a 3D structure. Source: PDB ID: 7DFR, Bystroff, C., Oatley, S. J., Kraut, J. (1990). Crystal structures of Escherichia coli dihydrofolate reductase: the NADP+ holoenzyme and the folate. NADP+ ternary complex. Substrate binding and a model for the transition state, *Biochemistry*, **29**, 3263-3277.

2.5.4 *Biomimetic polypeptides*

It is possible to design structures by mimicking proteins. Instead of using just the standard 20 amino acids, it is possible to use other molecules with similar properties and assemble them into chains. The manufacturing can utilize self-assembly techniques similar to those making up conventional polymers, and the RNA or DNA codons can be changed to correspond to other molecules than the standard 20 amino acids [43-46].

2.5.5 *Viruses*

Viruses are small containers of genetic information (RNA or DNA). They are autonomous genetic machines that reproduce themselves inside cells, and subsequently kill the cells that they infect. Because viruses are very small, difficult to be detected and highly capable of infecting a large number of people or animals, there is a threat for use of viruses as biological warfare agents. For example, polymyelitis viruses, which are among the smallest viruses, have a diameter about 30 nm. Bacteriophages are somewhat larger viruses. They have heads with a diameter about 40 nm and tails with dimensions 100 nm × 13 nm. It has been found that people can be infected by the bird flu H5N1. The HA associated with the virus makes it particularly lethal because people have no natural immunity to it. That is, their antibodies do not recognize this new HA antigen.

It is very interesting that C60 fullerenes and many viruses share similar morphologies. Fullerenes are also capable of penetrating the lipid bilayer of plasma membranes. In addition, fullerenes can be used as free radical generators under light irradiation where they initially select and target proteins and membranes [47-49]. Many of the similarities between fullerenes and viruses can be used in biology and medicine in both damaging and beneficial ways.

2.5.6 *Deoxyribonucleic acid*

Biology uses deoxyribonucleic acid (DNA) to store information. The whole blueprint of the structure of each cell in your body is stored in the DNA, and the information that tells certain cells to become liver, muscle or nerve cells, where and how to grow bones, which proteins to be manufactured and much else. DNA has so high information-carrying capacity that much of the information is probably redundant.

DNA is the encoder of genetic information. It consists of a sugar-phosphate backbone consisting of alternating groups of deoxyribose sugar and phosphate molecules. The genetic information is coded into four nucleotides. They can bind to each other with hydrogen bonds such that A always bonds to T and C always bonds to G. The overall structure is the familiar double helix. As already discussed before, DNA is a useful building block for nanostructures. One particular reason is that because the nucleotides can only pair with the exact opposite variety, one sequence of DNA is highly selective in which other DNA it can be attached to. The DNA chains can also be jointed with other molecules in the end, and this has allowed for making a branched DNA structure.

2.5.6.1 *Structures of DNA*

DNA stores and transmits genetic information, responsible for protein synthesis, and the ribonucleic acid (RNA) helps in protein synthesis. From the structure, RNA is different from DNA mainly in three ways. First, sugar in RNA is ribose, second, thiamine is replaced by uracil, and third, RNA is generally single stranded.

Out of the three groups of biological building blocks (lipids, proteins and DNA), the DNA molecule is chemically the simplest. Figure 2.21 shows the basic structure of DNA.

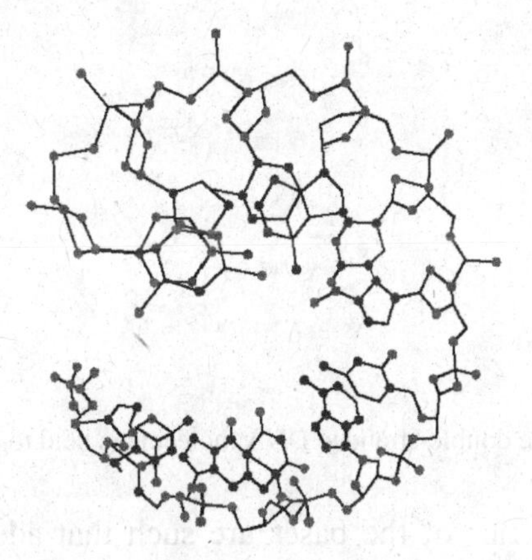

Figure 2.21 Basic structure of DNA.

DNA is made up of deoxyriboses, phosphates and a linear polymer composed of four subunits of nitrogen-containing compounds. Deoxyribose is a type of five-carbon sugar, and the four subunits are adenine, thymine, guanine and cytosine, which are more commonly represented by the letters A, T, G, C respectively. Figure 2.22 shows the structures of the deoxyribose and four subunits [4]. The four subunits are usually called four bases, and they are attached to the R group of the deoxyribose.

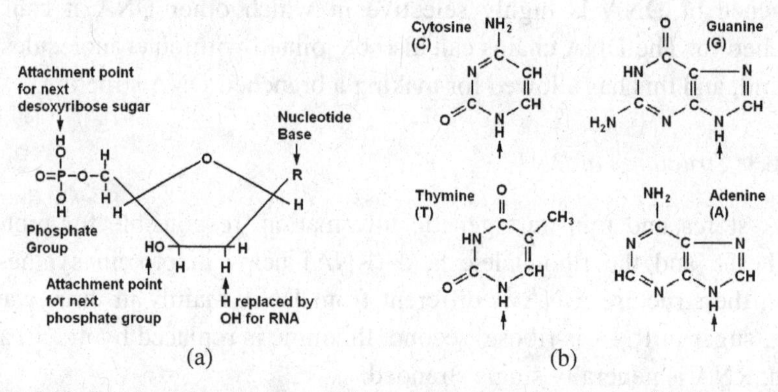

Figure 2.22 Sketch of the structures of deoxyribose (a) and four bases (b). Source: C. P. Poole, Jr. and F. J. Owens, Introduction to Nanotechnology, John Wiley and Sons, New Jersey. © 2003. Reprinted with permission of John Wiley and Sons, Inc.

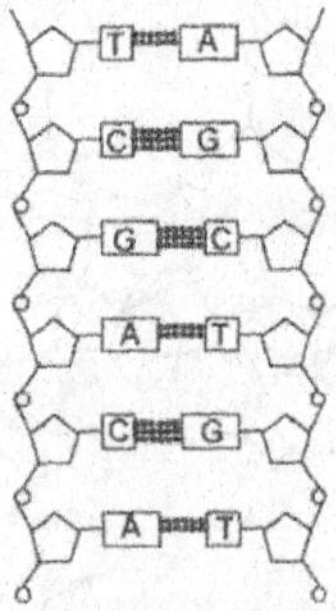

Figure 2.23 The double-stranded DNA molecule is held together by bases.

The characteristics of the bases are such that adenine (A) binds to thymine (T), and guanine (G) binds to cytosine (C), as shown in Figure 2.23. Therefore, two DNA polymers with the appropriate complementary sequences of nucleotides bind together, resulting in the famous double helix structure. The whole blueprint of the structure of each cell in your body is stored in the DNA.

In addition to double helix structure, it is possible to construct squares, cubes and even octahedral made from DNA. One DNA sequence can be designed to bind specifically to another DNA sequence. This makes DNA function very effectively as a kind of smart molecular glue for binding nanoscale objects. Although it is still a controversial area, a lot of research is going on to establish DNA's ability to act as a conductor, semiconductor or insulator depending on the DNA structure. The research on this area would add another important utility of DNA as a kind of nanomaterials.

2.5.6.2 *Nanoparticles and DNA*

To realize special functions, nanoparticles are often coated with organic molecules, and these organics can be DNA. Here we discuss DNA-gold-nanoparticle conjugates. Due to the surface plasmon resonance, monodisperse Au nanoparticles alone are red. Au nanoparticles in solution with DNA can form aggregates, and then are no longer red but purple in color. As shown in Figure 2.24, the DNA strands a′b′ (matching the a, b molecules on the surface of the nanoparticles) link them, shifting the plasmon resonance upwards in wavelength, from 520 to 574 nm.

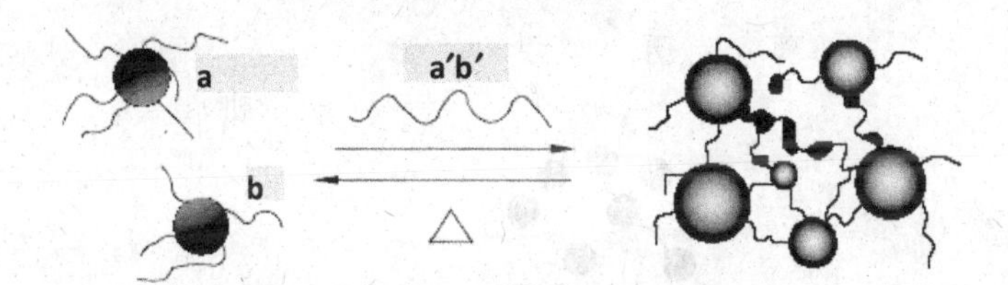

Figure 2.24 DNA-gold-nanoparticle conjugates.

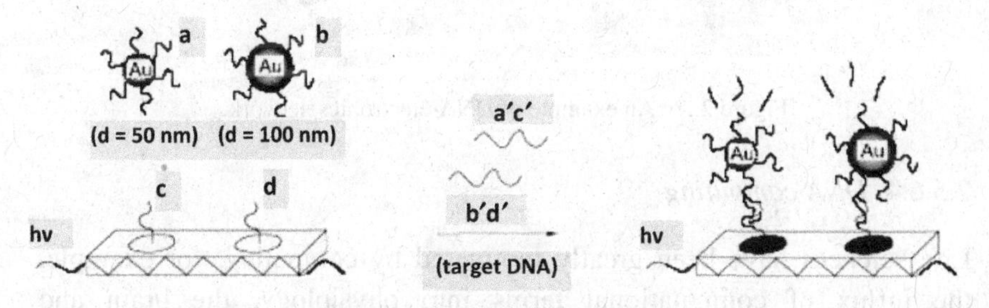

Figure 2.25 Detection of DNA.

This can be used in DNA detection systems. As shown in Figure 2.25, the nanoparticles can be bound to surfaces with DNA. When suitable DNA strands a′c′ or b′d′ are present which selectively match the a,c or b,d molecules, the nanoparticles will be bound to the surface. As the DNA-Au system is also electrically conductive, it may be used for electrical detection of DNA.

2.5.6.3 *DNA electronics*

DNA plays a crucial role in DNA electronics networks. As shown in Figure 2.26, the DNA network interwires the devices and connects them with electrode pads. Poly(dG)-poly(dC) has the best conductivity to act as a conducting nanowire. The electrode pads are short DNA molecules. The fabrication process involves two major steps. In the first step, assemble template or scaffold with molecular address defined by the DNA sequence and then add electronics functionality in DNA scaffold. In the second step, molecular devices are localized at specific addresses and the DNA molecules are converted into conducting wires which interconnect the devices and the electrodes.

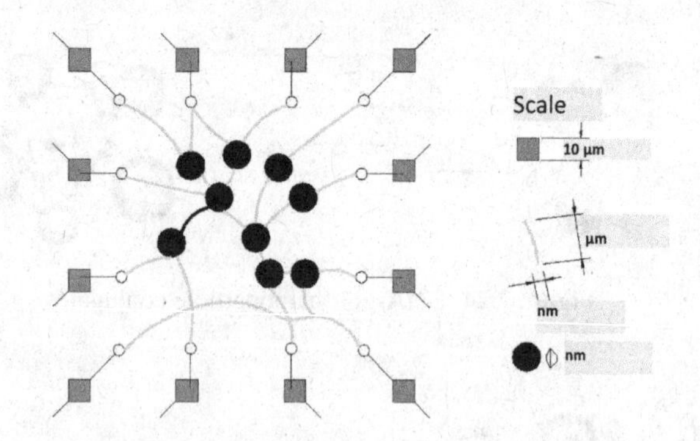

Figure 2.26 An example of DNA electronics network.

2.5.6.4 *DNA computing*

Life sciences have been greatly motivated by computing, for example, the influx of computational terms into physiology, the brain and cognitive sciences and evolutionary theory. The computing problems are getting far more complex than what were envisioned earlier. It seems that parallel computing is the answer to most of the complex problems. However, traditional microelectronics does not offer appropriate architectures and efficient wiring technologies for parallel processing, and these are the challenge for nanoelectronics. Bio-nanotechnology gives new concepts in the form of biochemical computers, such as the DNA computer and the quantum-mechanical computer. Although the realization of such computers is still far away, this is the scope of fast and efficient parallel computing beyond nanoelectronics.

2.5.7 *Biological neuronal networks*

The nerve cell, usually called neuron, is a functional unit of the body's nervous system that generates and transmits electrochemical impulses. These electrochemical impulses are the way by which information is exchanged in our bodies. If we take neurons as network lines that make up the Internet, electrochemical impulses are e-mail. Just as e-mail is used to send messages between people, electrochemical impulses are used to send messages between different body parts.

As shown in Figure 2.27, a single neuron is composed of a large cell body, the soma, and two cell parts: the dendrites and an axon. The dendrites send impulses to the soma while the axon sends impulses away from the soma. The axon terminates in a synapse that establishes the connection to other neurons. Dendrites, connected to the synapses are in charge of the input signals of the neuron.

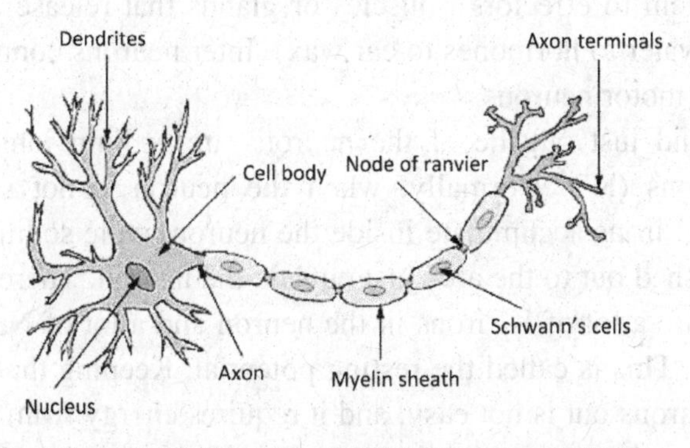

Figure 2.27 Structure of a biological neuron [54].

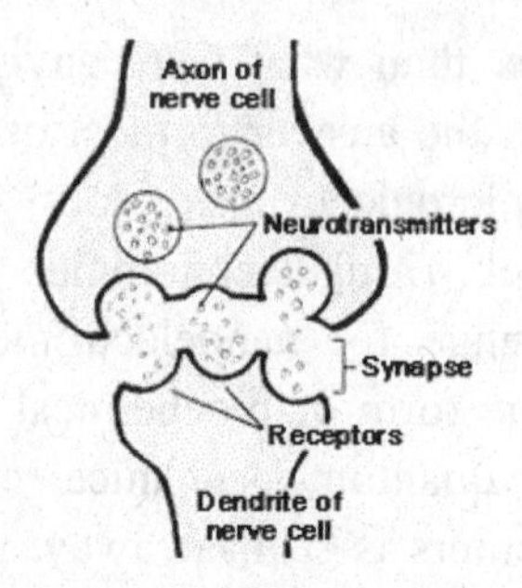

Figure 2.28 Basic structure of a synapse [55].

Figure 2.28 shows the basic structure of a synapse. Synapses can be excitatory or inhibitory and may change over time. When the inputs reach some threshold an action potential (electrical pulse) is sent along the axon to the outputs. The strength and polarity of the signal transmitted through a synapse are determined by the physical and neurochemical characteristics of the synapse. The amount of stimulation that the firing axon imparts to the neighboring dendrite can be changed by modifying the chemical constitutions of the neurotransmitters. Furthermore, changing the neurotransmitters can also change an excitatory stimulation to an inhibitory one.

According to their functions, neurons can be generally classified into three types: sensory neurons, motor neurons and inter neuróns. Sensory neurons carry signals from sense receptors (nerves that help us see hear, smell, taste and feel) to the central nervous system which includes the brain and the spinal cord. Motor neurons carry signals from the central nervous system to effectors (muscles or glands that release all kinds of stuff, from water to hormones to ear wax). Inter neurons connect sensory neurons and motor neurons.

Inside and just outside of the neurons are sodium ions (Na^+) and potassium ions (K^+). Normally, when the neuron is not sending any messages, K^+ irons accumulate inside the neuron while sodium ions Na^+ irons are pushed out to the area just outside the neuron. Therefore, at this state, there are a lot of K^+ irons in the neuron and a lot of Na^+ irons just outside of it. This is called the resting potential. Keeping the K^+ irons in and the Na^+ irons out is not easy, and it requires energy from the body to work. An impulse coming in from the dendrites, reverses this balance, causing K^+ irons to leave the neuron and Na^+ irons to come in. This is known as depolarization. As K^+ irons leave and Na^+ irons enter the neuron, energy is released, since the neuron is no longer doing any work to keep K^+ irons in and Na^+ irons out. This energy creates an electrical impulse or action potential that is transmitted from the soma to axon. As the impulse leaves the axon, the neuron re-polarizes: it takes K^+ irons back in and kicks Na^+ irons out and restores itself to resting potential. The neuron is then ready to send another impulse. The above process is very fast, and theoretically a neuron can send roughly 266 messages in one second. The electrical impulse may stimulate other neurons from its synaptic knobs to propagate messages.

2.5.8 *Artificial neuronal networks*

Inspired by the way in which biological nervous systems, such as the brain, various artificial neural networks (ANNs) have been developed for processing information [50, 51]. An ANN consists of a large number of highly interconnected processing elements (neurons) working in accord for solving specific problems. Usually an ANN is constructed for a specific application, such as data classification or pattern recognition, through a learning process. Like people, ANNs learn by example. In a biological system, learning involves the adjustments to the synaptic connections between the neurons, and an ANN does in the same way. A trained neural network is an "expert" in the category of information it has analyzed. The key advantages of an ANN mainly include:

- Adaptive learning: an ANN can learn how to do tasks based on previous experiences.
- Self-organization: an ANN can organize the information it receives during the learning time.
- Real time operation: an ANN can carry out computations in parallel.

2.5.8.1 *Neural networks and conventional computers*

A neural network takes a different approach for solving problem from that of a conventional computer. A conventional computer uses a cognitive approach for solving problems. The way for solving a certain problem must be stated in unambiguous instructions. These instructions should be converted to a high-level language program, which is further converted into machine code that could be understood by the computer. However, neural networks process information like the human brain does. Neural networks cannot be programmed to perform a specific task, but they can learn by examples. The examples should be carefully selected, otherwise time will be wasted, and even worse the network could not work correctly. One disadvantage of neural networks is that their operations could be unpredictable since they find out how to solve problems by themselves.

Biological neurons have relatively large dimensions compared to integrated CMOS circuits. They have dimensions in the order of millimeters, and the length of axons may be several centimeters and, in some cases, more than one meter. However, the three-dimensional structure of neuronal networks is very advantageous, in contrast to the presently limited two-dimensional integrated circuits. This limitation of current two-dimensional integrated circuits not only accounts for a low packing density, but also limits the wiring and architectural facilities.

Figure 2.29 shows an example of deriving an artificial neuronal network architecture from a biology-based neuronal network. Neurons can be treated in a first-order approximation as threshold gates. Two adjacent gate levels are fully interconnected. This architecture is equivalent to a matrix configuration. Each column corresponds to a neuron and each dot represents a synaptic coupling. In terms of artificial neurons, these couplings are referred to as weights, which evaluate the input signals.

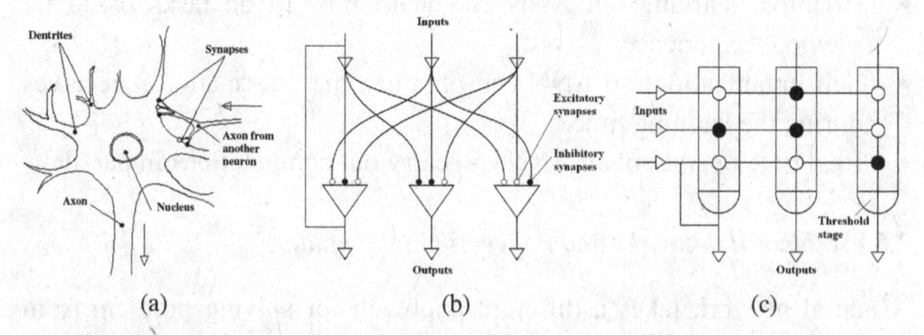

Figure 2.29 Derivation of a matrix architecture form a biological neuron. (a) Biological neuron, (b) threshold gate, (c) matrix architecture.

2.5.8.2 *Biological neuronal cells on silicon*

For the research on central nervous system and neurological implants, it is desirable that the growth and attachment of neurons can be controlled. The growth of neuronal cells on silicon and their combination with integrated circuits would have seemed technologically impossible a few years back. However, if the growth process takes place on top of a structured silicon substrate, the nerve fibers grow along these channels. According to this method, the shape of the neuronal network is determined by the silicon surface structure. MOS structures are capable of picking up the signals of the biological network, which enables the interconnection with integrated circuits [41]. As the packing density of this technology is very low at this stage, such concepts are not suitable for practical computer applications.

These concepts could be used for medical purposes, since the whole human body is accessible via its neuronal network. The information about the status of the different organs might be gathered by tapping the neuronal network, which could save the money spent for huge medical equipment. Today this is still a vision, but in the future it might be used in curbing the costs of the public health system.

In addition, these concepts are very interesting for the development of artificial limbs. They are also interesting for nanoelectronics because this hybrid technology combines biological neuronal networks with electrical circuits.

2.5.9 *Molecular biomimetics*

The term "biomimetics" was first introduced into the literature by Otto H. Schmitt, and his work helped us to get inspirations from biology for engineering solutions. In molecular biomimetics, hybrid technologies based on the tools of molecular biology and nanotechnology are used to mimic the nature.

Molecular biomimetics offers three solutions for the development of heterofunctional nanostructures. First, protein templates are designed at the molecular level through genetics. This ensures complete control over the molecular structure of the protein template, and actually this is a

DNA-based technology. Second, surface specific proteins can be used as linkers to bind synthetic entities, including nanoparticles, functional polymers or other nanostructures onto molecular templates, and this is a kind of molecular and nanoscale recognition. The third solution harnesses the ability of biological molecules to self- and co-assemble into ordered nanostructures, and this ensures a robust assembly process for achieving complex nanostructures.

However, it should be always kept in mind that although we are interested to mimic and to some extent imitate the nature, we do not want to slavishly copy the nature, therefore we use the term biomimetic nanotechnology.

References:

1. Feynman, R. (1960). There's plenty of room at the bottom, *Engineering and Science*, **23**, 22-36.
2. Wilson, M., Kannangara, K., Smith, G., Simmons, M. and Raguse, B. (2004). *Nanotechnology: Basic Science and Emerging Technologies*, Chapman and Hall/CRC.
3. Editorial, *Nature Nanotechnology*, **1** (2), 79, 2006.
4. Poole, C. P. and Owens, F. J. (2003). *Introduction to Nanotechnology*, Wiley Interscience.
5. Nalwa, H. S. (2000). *Handbook of Nanostructured Materials and Nanotechnology*, Volumes 1-5, Academic Press, Boston.
6. Cao, G. (2004). *Nanostructures and Nanomaterials*, Imperial College Press, London.
7. Kittel, C. (2004). *Introduction to Solid State Physics*, 8th ed., Wiley.
8. Losio, R. (2001). Band splitting for Si(557)-Au: Is it spin-charge separation?, *Physical Review Letters*, **86**, 4632-4635.
9. Robinson, I. K., Bennett, P. A. and Himpsel, F. J. (2002). Structure of quantum wires in Au/Si(557), *Physical Review Letters*, **88**, 096104.
10. Zettili, N. (2001). *Quantum Mechanics: Concepts and Applications*, Wiley, Chichester.
11. Crain, J. N., Kirakosian, A., Altmann, K. N., Bromberger, C., Erwin, S. C., McChesney, J. L., Lin, J. L. and Himpsel, F. J. (2003). Fractional band filling in an atomic chain structure, *Physical Review Letters*, **90**, 176805 (2003).
12. Zuhlehner, W. and Huber, D. (1982). *Czochralski Grown Silicon in Crystals*, Springer, Berlin.
13. Jasinski, J. M., Meyerson, B. S. and Scott, B. A. (1987). Mechanistic study of chemical vapor deposition, *Annual Review of Physical Chemistry*, **38**, 109-140.

14. McEllistrem, M., Allgeier, M. and Boland, J. J. (1998). Dangling bond dynamics on the silicon (100)-2x1 surface: Dissociation, diffusion, and recombination, *Science*, **279**, 545-548.
15. Zhang, Z., Wu, F. and Lagally, M. G. (1997). An atomistic view of Si(001) homoepitaxy, *Annual Review of Materials Science*, **27**, 525-553.
16. Tsuno, T., Imai, T., Nishibayashi, Y., Hamada, K. and Fujimori, N. (1991). Epitaxially grown diamond (001) 2x1/1x2 surface investigated by scanning tunneling microscopy in air, *Japanese Journal of Applied Physice Part 1 – Shot Notes and Review*, **30**, 1063-1066.
17. Davisson, C. and Germer, L. H. (1927). The scattering of electrons by a single crystal of nickel, *Nature*, **119**, 558-560.
18. Tromp, R. M., Hamers, R. J. and Demuth, J. E. (1986), Scanning tunneling of microscopy of Si(001), *Physical Review B*, **34**, 5343-5357.
19. Binnig, G., Rohrer, H., Gerber, C. and Weibel, E. (1983). 7x7 reconstruction on Si(111) resolved in real space, *Physical Review Letters*, **50**, 120-123.
20. Schlier, R. E. and Farnsworth, H. (1959). Structure and adsorption characteristics of clean surfaces of germanium and silicon, *Journal of Chemical Physics*, **30**, 917-926.
21. Tromp, R. M., Hames, R. J. and Demuth, J. E. (1985). Si(001) dimmer structure observed with scanning tunneling microscopy, *Physical Review Letters*, **55**, 1303-1306.
22. Tromp, R. M., Hamers, R. J. and Demuth, J. E. (1986), Scanning tunneling of microscopy of Si(001), *Physical Review B*, **34**, 5343-5357.
23. Christmann, K., Behm, R. J., Ertl, G., Van Hove, M. A. and Weinberg, W. H. (1979). Chemisorption geometry of hydrogen in Ni(111), *Journal of Chemical Physics*, **70**, 4168-4184.
24. Shih, H. D., Jona, F., Jepsen, D. W. and Marcus, P. M. (1976). Atomic underlayer formation during reaction of Ti[001] with nitrogen, *Surface Science*, **60**, 445-465.
25. Van Hove, M. A., Weinberg, W. H. and Chan, C. M. (1986). *Low-Energy Electron Diffraction*, Springer-Verlag, Berlin.
26. Finnis, M. W. and Heine, V. (1974). Theory of lattice contraction at aluminum surfaces, *Journal of Physics F – Metal Physics*, **4**(3), L37-L41.
27. Landman, U., Hill, R. N. and Mostoller, M. (1980). Lattice-relaxation at metal surfaces, *Physical Review B*, **21**, 448-457.
28. Adams, D. L., Nielsen, H. B., Andersen, J. N., Stengsgaard, I., Friedenhansl, R. and Sorensen, J. E. (1982). Oscillatory relaxation of the Cu(110) surface, *Physical Review Letters*, **49**, 669-672.
29. Chan, C. M., Van Hove, M. A. and Williams, E. D. (1980). R-factor analysis of several models of the reconstructed IR(110)-(1x2) surface, *Surface Science*, **91**, 440-448.
30. Robinson, I. K., Kuk, Y. and Feldman, L. C. (1984). Domain-structure of the clean reconstructed Au(110) surface, *Physical Review B*, **29**, 4762-4764.

31. Nutzenadel, C., Zuttel, A., Chartouni, D., Schmid, G. and Schlapbach, L. (2000). Critical size and surface effect of the hydrogen interaction of palladium clusters, *The European Physical Journal D*, **8**, 245-250.
32. Adamson, A. W. and Gast, A. P. (1997). *Physical Chemistry of Surfaces*, 6th edition, John Wiley and Sons, New York.
33. Goldstein, A. N., Echer, C. M. and Alivisatos, A. P. (1992). Melting in semiconducting nanocrystals, *Science*, **256**, 1425-1427.
34. Van Hove, M. A., Koestner, R. J., Stair, P. C., Birberian, J. P., Kesmodell, L. L., Bartos, I. and Somarjai, G. A. (1981). The surface reconstructions of the (100) crystal faces of iridium, platinum and gold. 1. Experomental observation and possible structural models, *Surface Science*, **103**, 189-217.
35. MacLaren, J. M., Pendry, J. B., Rous, P. J., Saldin, D. K., Somorjai, G. A., Van Hove, M. A. and Vvedensky, D. D. (1987). *Surface Crystallography Information Service*, Reidel Publishing, Dordrecht.
36. Herring, C. (1952). *Structure and Properties of Solid Surfaces*, University of Chicago, Chicago, IL.
37. Mullins, W. W. (1963). *Metal Surfaces: Structure Energetics and Kinetics*, The American-Society for Metals, Metals Park, OH.
38. Matijevic, E. (1985). Production of monodispered colloidal particles, *Annual Review of Materials Science*, **15**, 483-516.
39. Temperley, H. N. V. (1952). Statistical mechanics and the partition of numbers. 2. The form of crystal surfaces, *Proceedings of the Cambridge Philosophical Society*, **48**, 683-697.
40. Burton W. K. and Cabrera, N. (1949). Crystal growth and surface structure 1, *Discussions of the Faraday Society*, **5**, 33-39.
41. Goser, K., Glosenkotter, P. and Dienstuhl, J. (2004). *Nanoelectronics and Nanosystems*, Springer, Berlin.
42. Mader, S. S. (2001). *Biology*, McGraw Hill, Boston.
43. Ringsdorf, H., Schlarb, B. and Venzmer, J. (1988). Molecular architecture and function of polymeric orientated systems – models for the study of organization, surface recognition, and dynamics of biomemberanes, *Angewandte Chemie – International Edition*, **27**, 113-158.
44. Liu, J., Feng, X., Fryxell, G. E., Wang, L. Q., Kim, A. Y. and Gong, M. (1998). Hybrid mesoporous materials with functionalized monolayers, *Advanced Materials*, **10**, 161-65.
45. Pileni, M. P. (1998). Colloidal self-assemblies used as templates to control size, shape and self-organization of nanoparticles, *Supramolecular Science*, **5**, 321-329.
46. Nikoobakht, B., Wang, Z. L. and El-Sayed, M. A. (2000). Self-assembly of gold nanorods, *Journal of Physical Chemistry B*, **104**: 8635-8640.
47. Langevin, D. (1992). Micelles and microemulsions, *Annual Review of Physical Chemistry*, **43**, 341-369.

48. Mann, S. (2000). The chemistry of form, *Angewandte Chemie – International Edition*. **39**, 3392-3406.
49. Soten, I. and Ozin, G. A (1999). New directions in self-assembly: materials synthesis over 'all' length scales, *Current Opinion in Colloid and Interface Science*, **4**, 325-337.
50. Mead, C. (1989). *Analog VLSI and neural systems*, Addition-Wesley, New York.
51. Teal, G. K. (1976). Single-crystals of germanium and silicon – basic to transistor and integrated-circuit, *IEEE Transactions on Electron Devices*, **23**, 621-639.
52. Kulkarni, S. K. (2007). *Nanotechnology: Principles and Practices*, Capital Book Publishing Co., New Delhi, India.
53. Bystroff, C., Oatley, S. J., Kraut, J. (1990). Crystal structures of Escherichia coli dihydrofolate reductase: the NADP+ holoenzyme and the folate. NADP+ ternary complex. Substrate binding and a model for the transition state, *Biochemistry*, **29**, 3263-3277.
54. Web Books Publishing, http://www.web-books.com/eLibrary/Medicine/Physiology/Nervous/Nervous.htm.
55. National Institute on Drug Abuse, http://teens.drugabuse.gov/mom/mom_meth2.php.

Chapter 3

Nanoscale Fabrication and Characterization

In this chapter, we discuss typical techniques for the fabrication and characterization at the nanometer scale. These techniques provide the experimental bases for the research and development of nanoscience and nanotechnology.

3.1 Nanoscale Fabrication

Various fabrication processes have been developed and applied for nanoscale fabrication [1-4]. Usually, the selection of a fabrication process is based on the requirement and the type of nanomaterials to be fabricated. As shown in Figure 3.1, there are two general approaches for nanoscale fabrication: top-down and bottom-up.

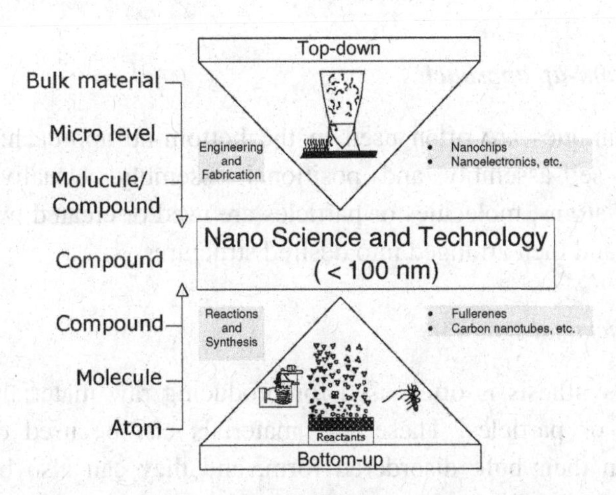

Figure 3.1 Top-down and bottom-up approaches for nanoscale fabrication.

The bottom-up approach has the following features. First, materials are synthesized from atomic or molecular species via chemical reactions. Second, chemical and biological reactors create conditions for special growth and assembly. Third, materials can self-assemble together to form more complex modules. Fourth, materials containing regular nano-sized pores can be used as templates for synthesizing nanoparticles, nanowires and nanotubes. Nanostructures fabricated with the bottom-up approach usually have less defects, more homogeneous chemical composition. The bottom-up approach is driven mainly by the reduction of Gibbs free energy, so that the nanomaterials produced are in a state closer to a thermodynamic equilibrium state.

The top-down approach has the following features. First, this approach breaks bulk materials into smaller sizes using lithography, mechanical and chemical manipulation tools. Second, optical, electron or ion beams are used for generating nanoscale patterns over a surface. Third, high precision actuators are used to move atoms or molecules from place to place. Fourth, micro tips are used to emboss or imprint materials. Top-down approach most likely introduces internal stress, in addition to surface defects and contaminations.

In the following, we discuss typical techniques for the bottom-up approach and top-down approach. More discussions about these two approaches can be found in Chapter 5.

3.1.1 *Bottom-up approach*

Three techniques are often used in the bottom-up approach: chemical synthesis, self-assembly and positional assembly. Usually, a large number of atoms, molecules or particles are used or created by chemical synthesis, and then arranged into desired structures.

3.1.1.1 *Chemical synthesis*

Chemical synthesis is often used for producing raw materials, such as molecules or particles. These raw materials can be used directly in products in their bulk disordered form, and they can also be used as building blocks for more advanced materials. In this technique, individual atoms or molecules are placed or self-assembled precisely to the places where they are needed. The chemical synthesis of various types of carbon nanomaterials is discussed in Chapter 4.

3.1.1.2 *Self-assembly*

In this technique, atoms or molecules arrange themselves into ordered nanoscale structures by physical or chemical interactions between the units. Typical examples of self-assembly include the formation of salt crystals and snowflakes with intricate structures. In some cases, an external force or field, for instance, electric or magnetic field, is applied to accelerate a slow self-assembly process. This technique is called directed self-assembly, and is often used in industries.

Based on different working mechanisms, different self-assembly processes have been developed, including chemical self-assembly, physical self-assembly and colloidal self-assembly. Chemical self-assembly converses molecular scale ordering of compounds with precisely designed atomic structures into more macroscopic structures. Physical self-assembly refers to the ordering of atoms that results from deposition processes, such as molecular beam epitaxy or chemical vapor deposition. By colloidal self-assembly, nanoparticles may aggregate into clusters.

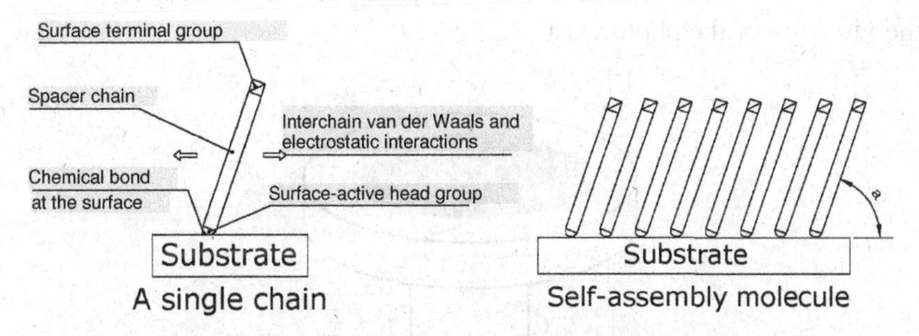

Figure 3.2 Building groups of self-assembly monolayers.

Self-assembly monolayers (SAM) are formed as a result of spontaneous self-organization of functionalized organic molecules onto the surfaces of appropriate substrates into stable, well-defined structures.

The final structure is close to or at thermodynamic equilibrium and it rejects defects. As shown in Figure 3.2, SAMs consist of three building groups: a head group (binding to a substrate), a surface terminal group, a spacer chain (backbone chain).

3.1.1.3 *Positional assembly*

In this technique, atoms, molecules or clusters are deliberately manipulated and positioned one-by-one. Usually, this technique needs scanning probe microscopes for working on surfaces, and optical tweezers for working in free space. At present, this technique is laborious, and is not suitable for industrial processes.

3.1.2 *Top-down approach*

The most popular top-down technique is nanolithography. In a conventional lithography process, materials are either deposited or removed on a substrate. The specific pattern developed on the substrate surface may lead to the changes in various properties, such as electrical, optical, etc., at that specific location depending the materials used [5, 6]. As shown in Figure 3.3, in a process of spin coating of photoresist layer, polymer goes on wet, and is then dried after spinning. The thickness of the material deposited depends on the spin speed of the spin coater and the viscosity of the photoresist.

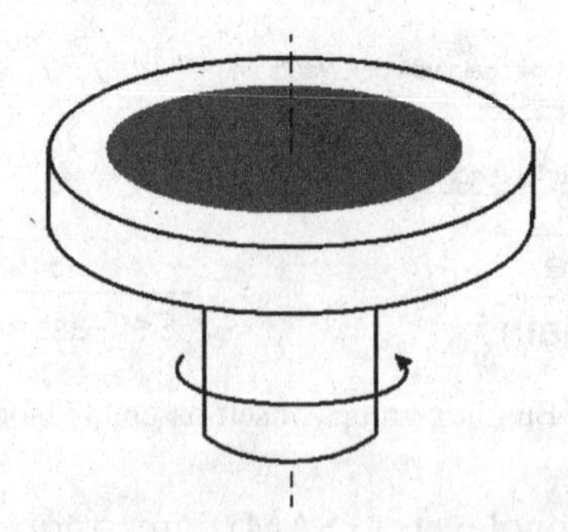

Figure 3.3 Spin coating of photo resist layer.

Nanolithography involves with the etching, writing or printing at the microscopic level on the order of nanometers. This technique may also be considered as a hybrid approach, since the growth of thin films is bottom-up whereas etching is top-down approach. The scanning probe microscopy (SPM) and the atomic force microscopy (AFM) are often used in nanolithography. By using the SPM, a surface can be viewed in fine details without modifying it. Both the SPM and the AFM can be used to etch, write or print on a surface in single-atom dimensions. In the following, we discuss typical lithography techniques.

3.1.2.1 *Photolithography*

As shown in Figure 3.4, photolithography, also called optical lithography, transfers a pattern from a photomask to the surface of a substrate. This technique is widely used for the fabrication of semi-conductor devices.

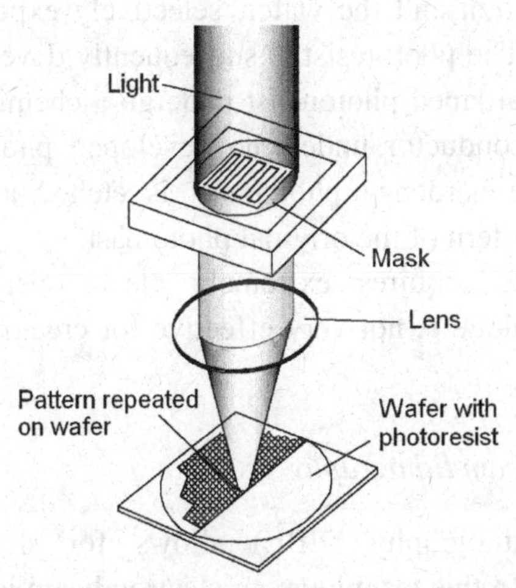

Figure 3.4 Schematic diagram showing a projection photolithography. (Image courtesy of Semiconductor Industry Association, with modifications made.)

Among various approaches for photolithography, projection lithography is the most commonly used. In this approach, the desired pattern is projected from the photomask onto the wafer. As shown in Figure 3.4, a light passes through a mask containing the desired image, and is then focused to produce the desired image on the wafer through a reduction lens system.

In this technique, the entire surface is drawn at a single moment, so this technique is a parallel method of nanolithography. Usually the substrates used are silicon wafers, and other options mainly include glass, sapphire and metal. However, if the wavelength of light used is extremely short, the lens may totally absorbs the light. This means that photolithography cannot reach the resolutions of some alternate technologies.

A full photolithography procedure involves a combination of substrate preparation, photoresist application, soft-baking, exposure, developing, hard-baking, etching and other chemical treatments in repeated steps on an initially flat substrate. A typical silicon lithography process would start with depositing a layer of conductive metal several nanometers thick on the substrate. Then a layer of photoresist is applied on top of the metal layer. After that, a photomask is placed between a source of illumination and the wafer, selectively exposing parts of the substrate to light. The photoresist is subsequently developed, and in this procedure the unhardened photoresist undergo a chemical change. After hard-baking, the conductor under the developed photoresist is etched away. Finally, the hardened photoresist is etched away, leaving the conductor in the pattern of the original photomask.

This technique requires extremely clean operation conditions. Besides, this technique is not very effective for creating shapes that are not flat.

3.1.2.2 *Electron beam lithography*

Electron beam lithography (EBL) allows for smaller sizes than photolithography. In this technique, an electron beam is used to generate patterns on a surface, and extremely small sizes on the order of nanometers can be achieved. This technique is one of the ways to beat the diffraction limit of light, and it has been used in industry, mainly for generating exposure masks to be used in conventional photolithography.

The task of EBL is to expose a pattern and remove, by etching or dissolution, the exposed portion of the polymer resist film. Due to the exposure, the molecular linear chains of the polymer on the exposed area are broken, and thus their average molecular weight is reduced. The reduction in the molecular weight of the polymer causes an increase of the solubility, which is related to the etching rate. Therefore a pattern drawn in terms of etching rate is obtained after the exposure, and it is desired that the contrast be as large as possible. The principle of EBL is shown schematically in Figure 3.5. This system has a high-speed electron gun which emits electrons, and the substrate is placed on a platform.

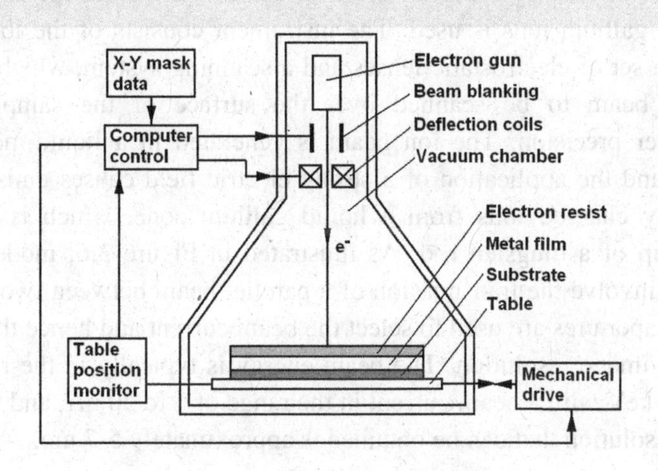

Figure 3.5 Schematic diagram of EBL process.

Based on current technology in electron optics, electron beam widths can go down to several nanometers, and this limit is mainly due to aberrations and space charge. However the practical resolution limit in BEL is mainly determined by forward scattering in the photoresist and secondary electron travel in the photoresist. Though the forward scattering could be decreased by using thinner photoresists and electron beams with higher energy, the generation of secondary electrons is inevitable.

Generally speaking, EBL is much more expensive and time consuming than photolithography. A job that takes one minute using photolithography would take more than one hour with electron-beam lithography.

3.1.2.3 *Focused ion beam lithography*

Combined with high-performance scanning electron microscopy (SEM), the focused ion beam (FIB) lithography technology is making a big impact on nanotechnology, particularly with the ability to use either focused ions or electrons to perform advanced nanolithography.

Figure 3.6 illustrates the basic operating principle of an FIB system. The focused ion beam instrument is similar in principle to an SEM. However, instead of using an electron beam to irradiate the sample, a beam of gallium ions is used. The instrument consists of the ion beam source, a set of electrostatic lenses and a scanning system, which allows the ion beam to be scanned over the surface of the sample with nanometer precision. The ion beam is generated in a liquid-metal ion source, and the application of a strong electric field causes emission of positively charged ions from a liquid gallium cone, which is formed on the tip of a tungsten rod. As illustrated in Figure 3.6, modern FIB systems involve the transmission of a parallel beam between two lenses. A set of apertures are used to select the beam current and hence the beam size and image resolution. The beam energy is typically in the range of 30 to 50 keV with a beam current in the range of 1 to 20 nA, and the best image resolution that can be obtained is approximately 5-7 nm.

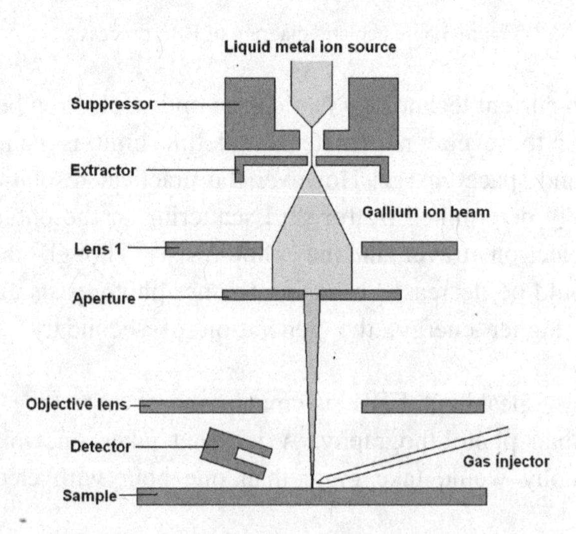

Figure 3.6 Basic principle of FIB equipment.

In the lithography process, the beam is raster-scanned over the sample, which is mounted in a vacuum chamber at a pressure of around 10^{-7} mbar. When the beam strikes the sample, secondary electrons and secondary ions are emitted from its surface. The electron or ion intensity is monitored and used to generate an image of the surface. Secondary electrons are generated in much greater quantities than ions and provide images of better quality and resolution. Therefore the secondary electron mode is used for most imaging applications.

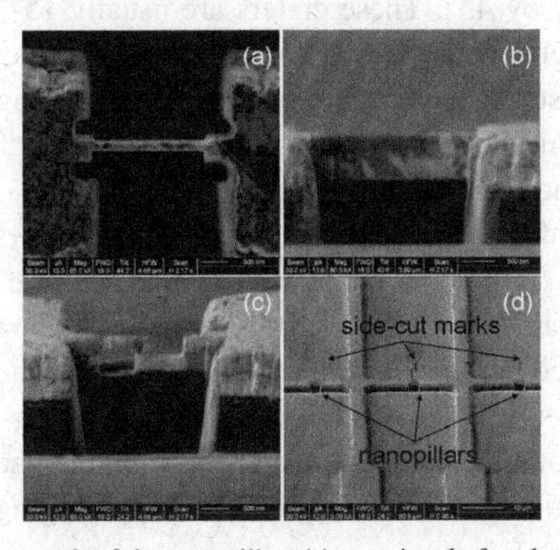

Figure 3.7 FIB micrograph of the nanopillar: (a) top view before lateral milling, (b) side view before lateral milling, (c) a completed nanopillar device and (d) low-magnification image of three nanopillars. Source: Wu, M. C., Aziz, A., Witt, J. D. S., Hickey, M. C., Ali, M., Marrows, C. H., Hickey, B. J. and Blamire, M. G. (2008). Structural and functional analysis of nanopillar spin electronic devices fabricated by 3D focused ion beam lithography, *Nanotechnology*, **19**, 485305. IOP Publishing Ltd.

Besides the conventional ion beam ligthography, three-dimensional (3D) focused ion beam lithography process has been developed, and used in the fabrication of 3D nanostructures and nanodevices [21]. Figure 3.7 shows a nanopillar fabricated by the 3D focused ion beam lithography process. This process exhibits two attractive advantages: firstly, this process preserves excellent interface cleanliness throughout the structure as all layers, including the electrodes, are deposited in a single ultrahigh vacuum process; secondly this process creates accurately rectangular devices rather than the more conventional circular or elliptical structures achieved by e-beam patterning. The former is important in eliminating the parasitic resistances and heating, and the latter enables accurate modeling of the device behavior [21].

Ion beams can also be used to remove material from the surface of the sample. This process, called milling, is a major advantage of FIB as much of the constructional analysis and failure analysis of semiconductor devices is performed on cross-sections. In a typical cross-sectional analysis, a crater is milled in the sample and the imaging is performed on the originally vertical wall of the crater after tilting the sample, generally by 45°. These craters are usually 15-20 μm wide and are milled in several steps. The initial crater has a stair case shape and is created using a strong beam current. The final milling of the wall is accomplished using line scans with a low beam current, so that the face obtained is flat and steep.

Similar to other analytical techniques, FIB analysis also has its drawbacks. Major problems include the damage to the milled surfaces from ion implantation, and the fact that some milling will occur during the imaging process. This milling slowly degrades the quality of the images. The latter problem can be avoided by using dual-beam FIB systems. These combine an FIB and an SEM column where the ion beam can be used for milling and the electron beam for imaging.

3.1.2.4 *Dip-pen nanolithography*

Dip-pen nanolithography (DPN) is a direct write lithographic technique that uses an AFM to build a pattern on the substrate material rather than etching it away. In the same way that an old fashioned dip-pen picks up ink from an ink well and then writes on a paper, in DPN, molecules are picked up from a reservoir on the end of the AFM tip and deposited to the surface of the substrate via a solvent or water as shown in Figure 3.8.

In order to create stable nanostructures, it is beneficial to use molecules that can anchor themselves to the substrate via chemisorption or electrostatic interactions. DPN can be used for both fabrication and imaging. This technique can be used to create patterns with resolutions down to 10 nm. Dip-pen nanolithography can be used in a range of applications from semiconductor patterning and chip manufacturing to biomedical and pharmaceuticals development.

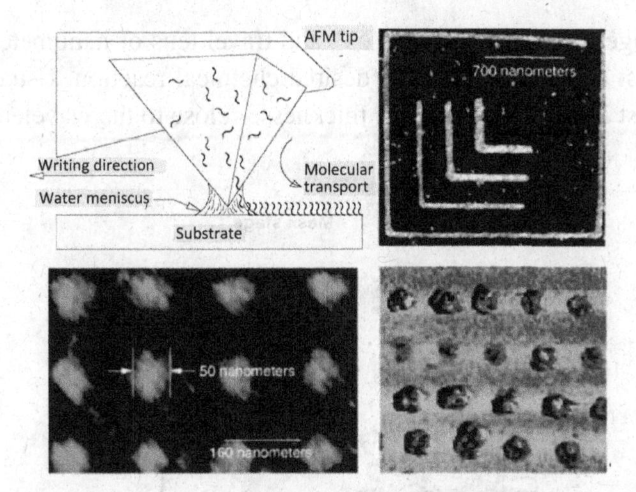

Figure 3.8 Schematic diagram showing how dip-pen nanolithography works. Several patterns obtained by DPN are also shown. Source: https://www.llnl.gov/str/December01/Orme.html.

3.1.2.5 *Extreme ultraviolet lithography*

Extreme ultraviolet lithography (EUVL) is similar to standard photolithography except that it uses intense beams of ultraviolet light reflected from a circuit design pattern, or mask, to burn the pattern into a silicon wafer [7]. It is an emerging contender for the replacement of optical photolithography in many areas. As all matters absorb extreme ultraviolet (EUV) radiations, EUVL should be performed in a vacuum. All the optical elements, including the photomask, must make use of defect-free Mo/Si multilayers which reflect light by means of interlayer interference, and each of these mirrors will absorb around 30% of the incident light. However, in a maskless interference lithography system, this limitation can be avoided, but such a system is restricted to producing periodic patterns only.

Figure 3.9 shows an EUVL apparatus [23]. In this system, the EUV is generated by the focused plasmas of laser or discharge pulses. The mirrors are responsible for collecting the light. When EUV photons are absorbed, photoelectrons and secondary electrons are generated by ionization, and this is similar to what happens when X-rays or electron beams are absorbed by a matter. These secondary electrons have energies in the range of a few to tens of eV, and travel tens of nanometers inside photoresist before initiating the desired chemical reaction. Usually EUV photoresist images require resist thicknesses close to the wavelength.

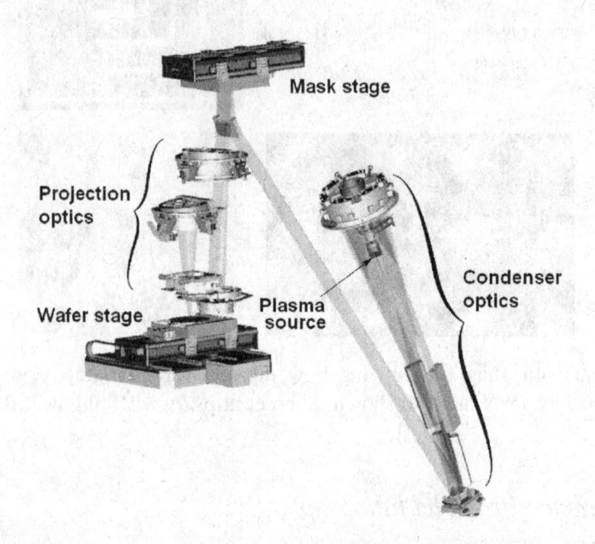

Figure 3.9 The optical layout of the engineering test stand for EUVL. Source: Sweeney, D. (1999). Extreme ultraviolet lithography: imaging the future, Science and Technology Review, November, 4-9. (Image courtesy of Lawrence Livermore National Laboratory, U.S. Department of Energy's National Nuclear Security Administration).

3.1.2.6 *Nanoimprint lithography*

Nanoimprint lithography is a novel method for the fabrication of nano scale patterns [8-10]. It is a simple process with high resolution, high throughput and low cost. The working principle of nanoimprint lithography (NIL) is similar to a rubber stamp. In this technique, a rubber-like polymer is inscribed a pattern, and its surface is coated with molecular ink. Then the ink is stamped out onto a surface, such as metal, polymer, oxide and other surface. In this process, patterns can be imprinted on micro and nano levels. It has a variety of applications such as single electron transistor device fabrication, fluidic channel fabrication, bio-molecule patterning for biosensors, polymeric material processing, pattering organic semiconductors for OLED and FET and polymer based MEMS and NEMS device.

There are two types of nanoimprint lithography processes: thermoplastic nanoimprint lithography and photo nanoimprint lithography.

(1) *Thermoplastic nanoimprint lithography*

In a thermoplastic nanoimprint lithography (T-NIL) process, a thin layer of imprint resist (thermoplastic polymer) is spin coated onto the sample substrate. In the procedure of T-NIL shown in Figure 3.10, a layer of prepolymer is deposited over the master. By heating up the polymer above its glass transition temperature, the pattern on the master is pressed into the melt polymer film [11]. After polymerization, an elastomeric stamp is obtained. The printing is then carried out using ink solution. A pattern transfer process, for instance, reactive ion etching, can be used to transfer the pattern in the resist to the underneath substrate.

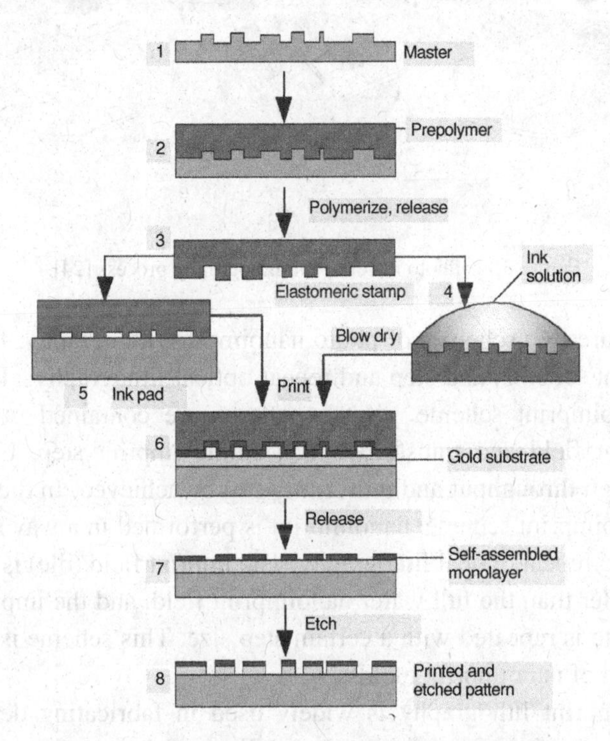

Figure 3.10 Procedure of nanoimprint lithography [11].

(2) *Photo nanoimprint lithography*

In a photo nanoimprint lithography (P-NIL) process, a photo (UV) curable liquid resist is applied to the substrate, and usually the mold is made of transparent material, such as fused silica. After the mold and the substrate are pressed together, the liquid resist is cured by UV light and becomes solid. After the mold is separated, the pattern in resist is transferred onto the underneath material by a pattern transfer process. Usually the resist is a monomer or polymer which is cured by UV light during the imprinting procedure, and the adhesion between the resist and the mold is controlled to ensure proper release. This process is schematically shown in Figure 3.11.

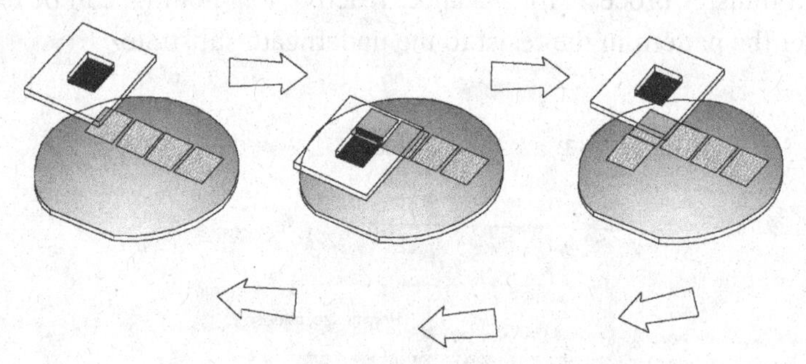

Figure 3.11 Photo nanoimprint lithography process [24].

There are two schemes of photo nanoimprint lithography: full wafer nanoimprint scheme, and step and repeat optical lithography. In the full wafer nanoimprint scheme, all the patterns are contained in a single nanoimprint field and transferred in a single imprint step. Using this scheme, high throughput and uniformity can be achieved. In the step and repeat nanoimprint scheme, nanoimprint is performed in a way similar to the step and repeat optical lithography. The imprint field (die) is typically much smaller than the full wafer nanoimprint field, and the imprinting to the substrate is repeated with a certain step size. This scheme is ideal for the creation of nanoimprint molds.

Nanoimprint lithography is widely used in fabricating devices for various applications. In electronics, NIL can be used to fabricate MOSFET, O-TFT, single electron memory. In photonics and optics, this technique can be used in the fabrication of subwavelength resonant grating filter, polarizers, anti-reflective structures, integrated photonics circuit and so on. In biology, NIL technique has been used to fabricate sub-10 nm nanofluidic channels that can be used in DNA diagnosis, and this technique can also be used in miniaturizing various biomolecular sorting devices.

Nanoimprint lithography is a simple and inexpensive technique. This technique does not require complex optics or high-energy radiation sources. There is no need for photoresists finely tailored for resolution and sensitivity at a specified wavelength. Furthermore, since large areas can be imprinted in one step, this is also a high-throughput technique.

3.1.2.7 *Contact lithography*

This technique is also called contact printing. In this technique, the image to be printed is obtained by illumination of a photomask which is in direct contact with a substrate coated with a layer of photoresist. This technique is a candidate for sub-45 nm semiconductor lithography.

The chief advantage of contact lithography is the elimination of the need for complex projection optics between object and image. The resolution limit in today's projection optical systems originates from the finite size of the final imaging lens and its distance from the image plane. More specifically, the projection optics can only capture a limited spatial frequency spectrum from the object (photomask). Contact printing has no such resolution limit but is sensitive to the presence of defects on the mask or on the substrate.

However, due to the photomask-photoresist interface, the image-forming light is subject to near-field diffraction as it propagates through the photoresist. Because of the diffraction, the image contrast will decrease with increasing depth into the photoresist. By using a thinner photoresist, this effect can be partly alleviated.

3.1.2.8 *X-ray lithography*

X-ray lithography is a next generation lithography developed for the semiconductor industry [12-14]. In this technique, collimated X-rays are used as the exposing energy. The short wavelengths of X-rays overcome the diffraction limits in the resolution of conventional optical lithography. As shown in Figure 3.12, X-rays illuminate a mask placed near to a wafer coated with a layer of resist, and no lenses are used. The mask used in X-ray lithography consists of an X-ray absorber on a membrane that is transparent to X-rays [15]. Typically, the material for the X-ray absorber is gold or compounds of tantalum or tungsten, and the material for the membrane is silicon carbide or diamond. The pattern on the mask is written by electron beam lithography onto the resist coated on the wafer.

Typically, the X-rays used in this technique are generated by a compact synchrotron radiation source, and their wavelengths are around 0.8 nm. To perform deep X-ray lithography, X-rays with shorter wavelengths, about 0.1 nm, are needed. With modified procedures, deep X-ray lithography can be used to fabricate deeper structures, even three-dimensional structures.

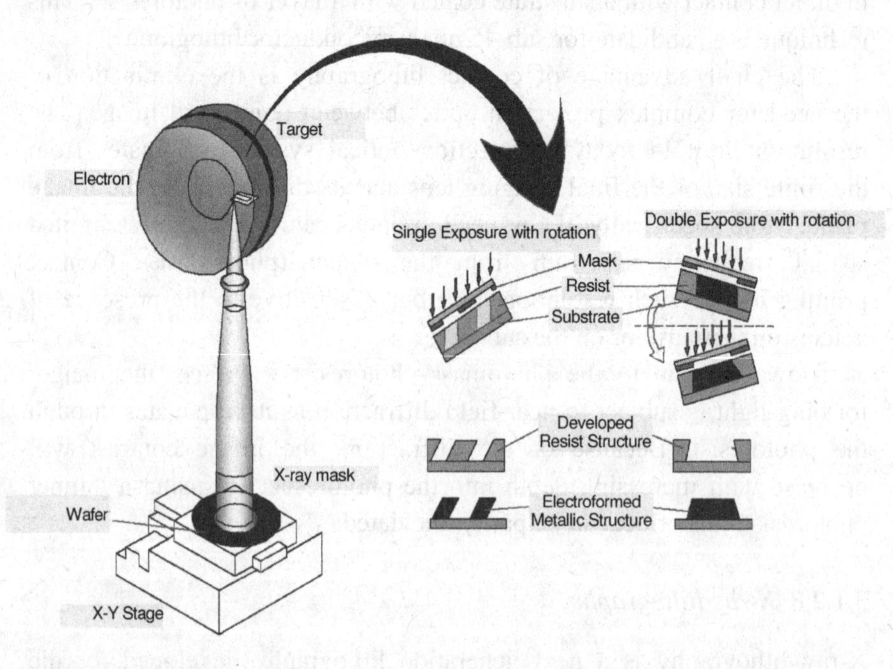

Figure 3.12 Schematic view of X-ray lithography (MIRRIRCLE – 20SX, Photon Production Laboratory). [25]

3.1.2.9 *Lift-off lithography*

Lift-off lithography is a technique to fabricate nanostructures by taking advantage of the AFM tip sharpness. A narrow furrow is engraved in a soft polyimide layer. The furrow is then transferred using dry etching to a thin germanium layer which forms a suspended mask. Metallic layers are then evaporated through this mask. Using this technique, metallic lines with line width in nanometers and single-electron transistors can be fabricated. Figure 3.13 schematically shows the typical process of lift-off lithography [22].

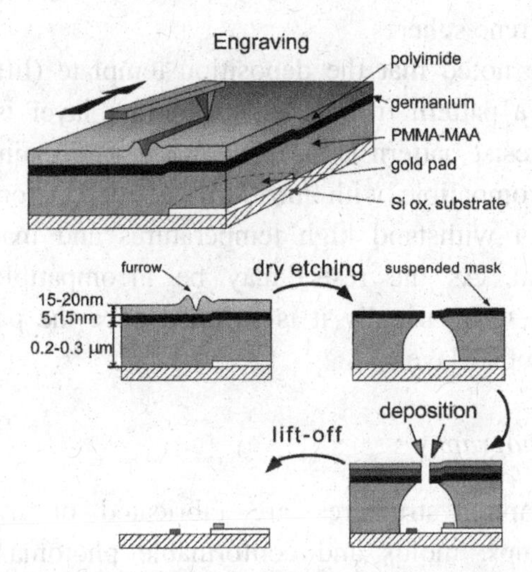

Figure 3.13 Main steps of the AFM-based trilayer process. Reprinted with permission from: Bouchiat, V. and Esteve, D. (1996). Lift-off lithography using an atomic force microscope, *Applied Physics Letters*, **69** (20), 3098-3100. © 1996, American Institute of Physics.

This technique can be used on any substrates and allows easy alignment with previously fabricated structures. One attractive feature of this technique is that many sorts of surfaces and paints (metals and molecules) can be used, and several layers of paint can be put down sequentially.

The lift-off technique can be used to produce monolayers. The fabrication of monolayers is based on the capillary force (surface tension) due to meniscus formation and the convective flow due to water

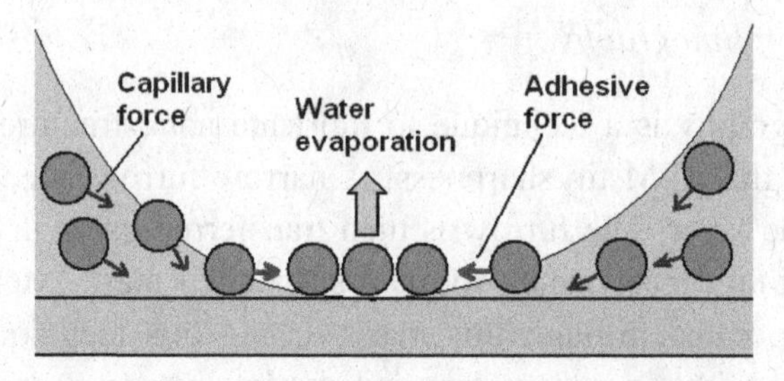

Figure 3.14 Fabrication of monolayers of polystyrene spheres [26].

evaporation. Figure 3.14 schematically shows the fabrication of monolayers of polystyrene spheres.

It should be noted that the deposition template (lift-off) approach for transferring a pattern from resist to another layer is less common than using the resist pattern as an etch mask. The main reason is that the resist is incompatible with most MEMS deposition processes, as it usually cannot withstand high temperatures and may be a source of contamination. As the resist may be incompatible with further micromachining steps, usually it is stripped once the pattern has been transferred to another layer.

3.1.2.10 *Soft lithography*

In soft lithography, structures are fabricated or replicated using elastomeric stamps, molds and conformable photomasks. The word "soft" indicates that elastomeric materials are used. This technique is often used in constructing features measured on the nanometer scale. Soft lithography includes the technologies of microcontact printing, replica molding, microtransfer molding, micromolding in capillaries and solvent-assisted micromolding.

In a typical procedure of soft lithography as schematically shown in Figure 3.15, the photoresist is applied on a clean wafer and then the master is developed using UV. This leads to the development of polydimethylsiloxane (PDMS) master and mould. PDMS is a type of soft silicone polymer. Polylactic-co-glycolic acid (PLGA) is applied over the PDMS mould, and finally PLGA of desired size and shape is obtained.

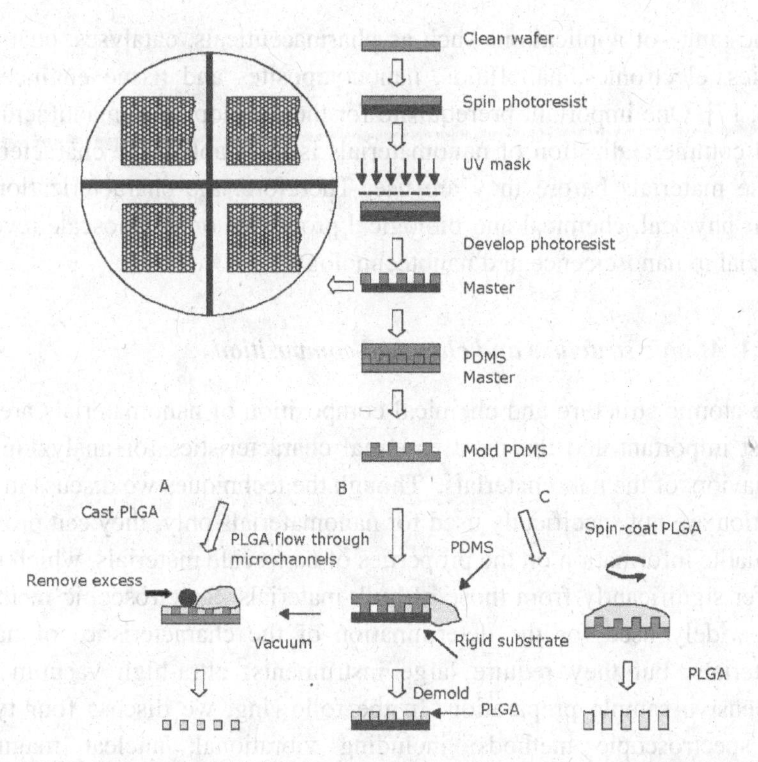

Figure 3.15 Procedure of soft lithography. Image courtesy of Center E. Piaggio [27].

This technique exhibits obvious advantages over other forms of lithography, such as photolithography and electron beam lithography. Its cost is lower than that of traditional photolithography in mass production. It does not need a photo-reactive surface to create a nano-structure, and more pattern-transferring technologies can be used in this technique. It has higher resolution than photolithography in laboratory settings (~30 nm versus ~100 nm). This technique is well-suited for applications in plastic electronics and biotechnology. Furthermore, it is well-suited for applications involving large or nonplanar surfaces.

3.2 Characterization of Nanomaterials

Characterization of nanomaterials is an area of great scientific interest as bottom-up approaches are undergoing exponential growth to fulfill a wide range of applications such as pharmaceuticals, catalyses, coatings, optics, electronics, nanofluids, nanocomposites and tissue engineering [16, 17]. One important prerequisite for the development, manufacturing and commercialization of nanomaterials is to establish the characters of these materials before they are use. Therefore, the characterization of their physical, chemical and biological properties on a nanoscale level is crucial in nanoscicence and nanotechnology.

3.2.1 *Atomic structure and chemical composition*

The atomic structure and chemical composition of nanomaterials are the most important and the most essential characteristics for analyzing the behaviors of the nanomaterials. Though the techniques we discuss in this section are not specifically used for nanomaterials only, they can provide valuable information on the properties of nanoscale materials, which may differ significantly from those of bulk materials. Spectroscopic methods are widely used for the determination of the characteristics of nanomaterials, but they require large instruments, ultra-high vacuum and expensive sample preparation. In the following, we discuss four types of spectroscopic methods, including vibrational, nuclear magnetic resonance, X-ray and UV spectroscopies, which have been extensively used for the characterization of nanomaterials.

3.2.1.1 *Vibrational spectroscopies*

Vibrational spectroscopies comprise Fourier transform infrared (FTIR) spectroscopy and Raman scattering (RS). These two methods can be used to investigate the vibrational structures of molecules or solids. FTIR is more suitable for organic compounds and used extensively for them. Both of the methods can be performed on dry powders or liquid suspensions. For FTIR, the absorption spectra are deduced from transmission measurements through a KBr pellet with entrapped nanoparticles or directly on nanoparticles in a reflection mode measurement (DRIFT). FTIR can be also used to determine the crystallization and grain sizes of nanostructured powders and bulk materials. The principle of FTIR spectroscopy is schematically shown in Figure 3.16. In FTIR,

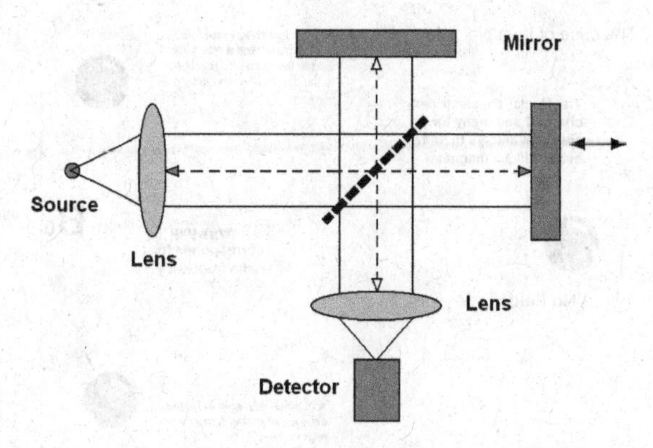

Figure 3.16 Principle of FTIR spectroscopy.

the light source is IR and the response of IR after passing through the specimen is detected by an IR detector.

3.2.1.2 *Nuclear magnetic resonance*

High resolution liquid and solid state nuclear magnetic resonance (NMR) is an important tool widely used for the characterization of nanomaterials. It can be used in identifying the elements, types of bonds, nanoscale effects, and the surface properties and chemistry of nanolayer systems. Figure 3.17 shows the working principle of NMR. Due to its spin, an electrically charged nucleus behaves like a magnet. When this nucleus is bombarded on the sample under an external magnetic field, magnetic resonance occurs. The resonance peak detected by the detector gives the information of the elements present in the nanomaterial.

3.2.1.3 *X-ray and UV spectroscopies*

X-ray and UV spectroscopies can be used to investigate the electronic structures of nanomaterials, from which their atomic structures can be deduced, such as core levels, valence and conduction band. X-ray photoemission spectroscopy (XPS) or electron spectroscopy for chemical analysis (ESCA) can be used to study the photoemission of electrons produced by a monochromatic X-ray or UV beam as shown in Figure 3.18.

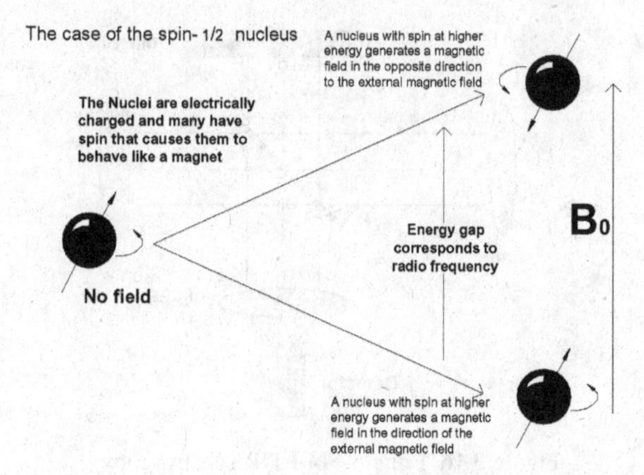

Figure 3.17 Schematic view of NMR spectroscopy. Source: Roy Hoffman and Yair Ozery, http://chem.ch.huji.ac.il/nmr/whatisnmr/whatisnmr.html. This diagram is the property of the Hebrew University of Jerusalem.

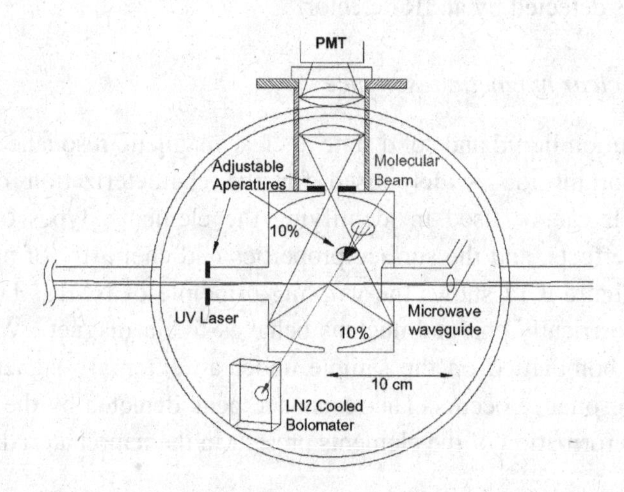

Figure 3.18 Principle of UV spectroscopy [28].

XPS techniques can be used to measure the kinetic energies of electrons. Due to the limited mean free path of the electrons in a matter, only few nanometric layers can be investigated. Since binding energies are highly sensitive to chemical bondings, a map of the bonding configuration can be obtained for surface layers. With the photoemission cross-sections, chemical compositions of the surface material can be calculated and compared to bulk chemical compositions.

X-ray absorption spectroscopy (XAS) is another X-ray method that can be used to investigate the conduction band of materials. This method involves extended X-ray absorption fine structure (EXAFS) and X-ray absorption near edge structure (XANES). The principle is based on the absorption of a monochromatic X-ray beam by a core shell electron of selected atomic species inside a sample. XAS techniques can be adapted for samples with low crystallinity. These local order techniques can be used to follow the early stages of the crystallization of amorphous nanoparticles. However, an important drawback is that XAS techniques usually require synchrotron radiation facilities.

3.2.1.4 *X-ray and neutron diffraction*

Diffraction techniques are often used to characterize the atomic structures of crystals [18]. Theoretically it is a diffraction of an incident beam (X-ray or neutrons) by the reticular planes of the crystalline phases inside a sample. As shown in Figure 3.19, the beam is diffracted at specific angular positions with respect to the incident beam depending on the phases of the sample. When the crystal size is reduced toward the nanometer scale, the diffraction peaks are broadened, and the widths of the peaks are directly correlated to the size of the nanocrystalline domains.

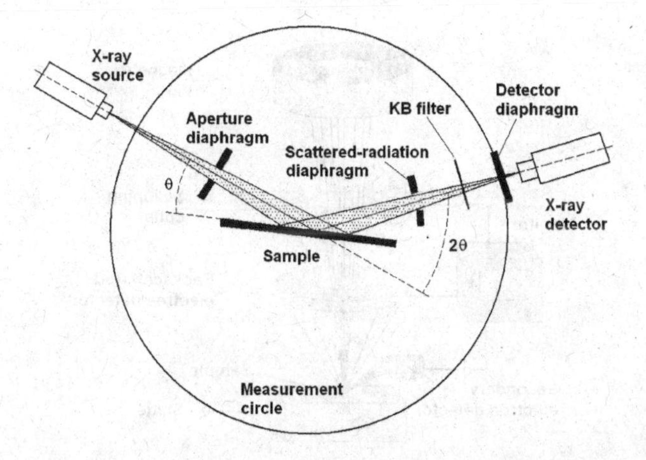

Figure 3.19 Principle of X-ray diffraction technique [29].

The X-ray and neutron diffraction techniques are extensively used for the characterization of various types of porous materials, providing quantitative parameters such as pore size, surface area and pore volume. In addition, they also allow the determination of the shape and, in particular, the spatial distribution of the pores.

3.2.2 *Size, shape and surface area*

In the research of nanoscience and nanotechnology, it is very important to characterize the size, shape and surface area of nanomaterials. In the following we discuss typical techniques for the characterization of the size, shape and surface area of nanomaterials.

3.2.2.1 *Electron microscopy*

Electron microscopies, mainly including scanning electron microscopy (SEM) and transmission electron microscopy (TEM), are powerful methods often used to investigate the properties of materials, such as

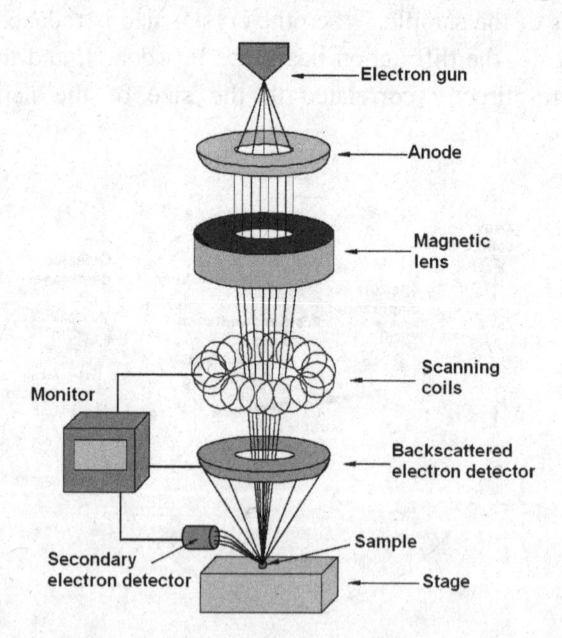

Figure 3.20 Working principle of SEM [30].

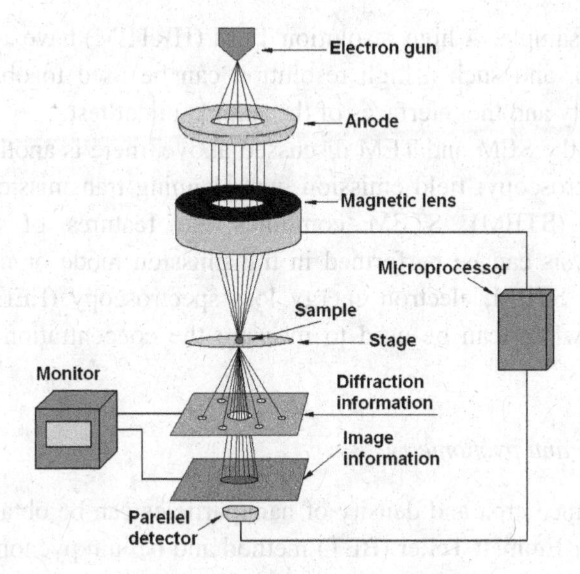

Figure 3.21 Principle of TEM [30].

size, shape and chemical composition, but the preparations of the samples for these methods are complicated. As shown in Figure 3.20, in an SEM system, electrons are produced from an electron gun, and then accelerated through anode plate and focused by magnetic lens [19]. The scanning coils force the electron beam to rapidly scan over an area of the specimen. Electrons emitted from a filament are reflected by the sample and images are formed using either secondary electrons or backscattered electrons. To investigate the nanometric scale, field emission microscope (FE-SEM) is required. FE microscopes could reach resolutions of the order of 1 nm using a cold cathode. If they are equipped with an energy dispersive spectrometer (EDS), the size distribution, shape and chemical composition of nanoparticles can be investigated by FE-SEM.

In TEM experiments, electrons pass through the sample under test and the transmitted beam is used to build the images, from which both the surface and the crystallographic information of the sample can be obtained. As shown in Figure 3.21, in a TEM system, electrons are produced from an electron gun, and then accelerated through an anode plate and focused by magnetic lens. The specimen should be very thin preferably in the order of 0.1 micrometer, so that electrons can pass through the sample. A high resolution TEM (HRTEM) have a resolution below 1 nm, and such a high resolution can be used to observed the crystal quality and the interfaces of the sample under test.

Besides the SEM and TEM discussed above, there is another kind of electron microscopy: field emission gun scanning transmission electron microscopy (STEM). STEM combines the features of SEM and TEM. Analysis can be performed in transmission mode or in scanning mode. With STEM, electron energy loss spectroscopy (EELS) can be performed, which can be used to measure the concentration profile of nanoparticles.

3.2.2.2 *BET and pycnometry*

Specific surface area and density of nanoparticles can be obtained using the Brunauer Emmett Teller (BET) method and helium pycnometry [20]. BET and He pycnometry are performed on dry powders. Helium pycnometry is a method for measuring the true bulk density of particles if they do not contain closed pores. In helium pycnometry, the variation of the helium pressure produced by a variation of volume is studied. BET is based on the measurement of the adsorption isotherm of an inert gas at the surface of the particles. In BET, Helium and nitrogen gas are fed to chamber in which sample is placed as shown in Figure 3.22 [31]. The change in temperature is recorded by data acquisition system and the results are analyzed and extrapolated to calculate the density and surface area of the specimen.

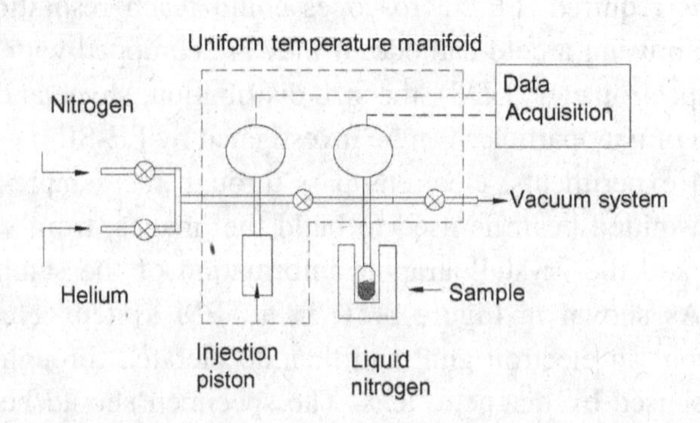

Figure 3.22 Working principle of BET method [31].

3.2.2.3 *Epiphaniometer*

An epiphaniometer is an instrument that can be used to measure the surface concentration of aerosol particles in both the nuclei and accumulation mode, and it is most sensitive to particles in the accumulation mode. Low atmospheric particle concentrations in remote locations can also be effectively measured using this method.

In this method, aerosol is passed through a charging chamber, and in the chamber, the lead isotopes created from a decaying actinium source are attached to the particle surfaces. The particles are transported through a capillary to a collecting filter. The radioactivity level of the particles collected on the filter is measured by a surface barrier detector.

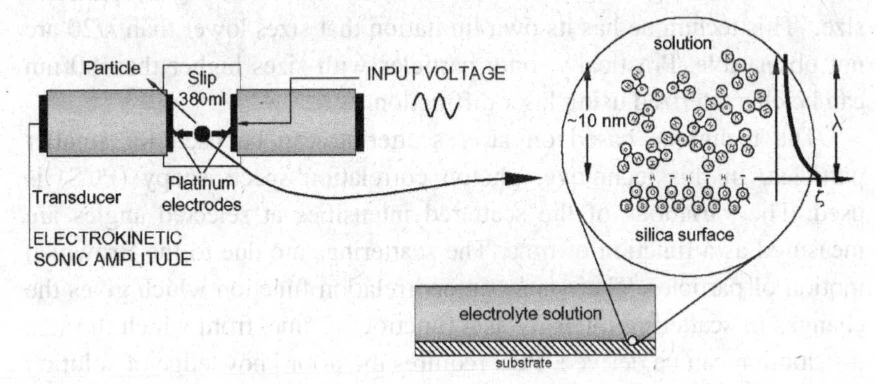

Figure 3.23 Principle of Zeta potential. Sources: Brian Kirby, Cornell University, http://www.kirbyresearch.com; and Zeta Potential, Dr. Giuliano Tari, CERAM.

3.2.2.4 *Zeta potential analyzer*

Zeta potential is a measurement of the charges carried by nanoparticles in suspensions. It is important that the liquid used should have requisite viscosity to keep nanoparticles suspended in the measurement duration. The principle of the commonly used Zeta potential analyzer is based on electrophoresis. As shown in Figure 3.23, the sample is placed between two transducers and a voltage is applied. The change in output gives a measurement of Zeta potential of the sample. Zeta potential measurements are often used to characterize the stability of the suspensions with electrostatic repulsion. Laser granulometries, which will be discussed below, are frequently used in combination with Zeta potential analyzers.

3.2.2.5 *Laser granulometry*

Laser granulometry is a statistical method for the determination of quantitative particle size distributions. Generally speaking, there are two techniques for laser granulometry. One is based on the diffraction of a laser beam by particles in stable suspensions, and the other is based on the scattering of a laser beam by particles in stable suspensions.

In the technique based on laser diffraction, the diffraction pattern (width of the ring and intensity) is directly connected to the particles size. This technique has its own limitation that sizes lower than $\lambda/20$ are not observable. Practically, only particles with sizes higher than 80 nm can be characterized using laser diffraction.

The technique based on laser scattering can be used for smaller particles. In this technique, photon correlation spectroscopy (PCS) is used. The variations of the scattered intensities at selected angles are measured as a function of time. The scatterings are due to the Brownian motion of particles. There is an autocorrelation function which gives the changes of scattering intensity as a function of time, from which the size distribution can be derived. PCS requires the prior knowledge of solution viscosity and refractive index, and is highly sensitive to the presence of agglomerates.

3.2.2.6 *Elliptically polarized light scattering*

Based on laser light scattering, this method can be used to investigate the size distribution and the shape distribution of nanoparticles, and it also can be used to study the structure and size distribution of agglomerates. In this method, incident laser beam is elliptically polarized, and the modifications of the polarization state due to the sample are measured at specific angular positions. Polarization analysis gives complementary information about the size and structure of agglomerates. In the field of nanomaterials research and development, many properties depend upon the agglomeration of nanoparticles. Agglomerates with size in the range between 50 nm and 2 μm can be characterized by this method.

3.2.2.7 *Gas adsorption*

A material possessing just one type of pore, even when the pores are disordered, might be more homogeneous than one having just a fraction of nicely ordered pores. Adsorption analysis is helpful for deciding and classifying porous materials according to the size of their pores. The correlation between the vapor pressure and the pore size is given by the Kelvin equation:

$$r_p\left(\frac{p}{p_0}\right)=\left(\frac{2\gamma V_L}{RT\ln\left(\frac{p}{p_0}\right)}\right)+t\left(\frac{p}{p_0}\right) \quad (3.1)$$

where r_p is the pore radius, p_0 is the initial pressure, p is the final pressure, γ is the surface tension, t is the thickness of the adsorbate film, R is the Ritzberg constant, T is the temperature and V_L is the molecular volume of the condensate.

Gas adsorption method is widely used for characterizing micro and mesoporous materials, and provides porosity parameters such as pore size distributions, surface areas and pore volumes. In a typical adsorption experiment, the uptake of gases, such as nitrogen, krypton and CO_2, is measured as a function of relative pressures $p/p_0 < 1$ at a constant temperature, where p and p_0 are the equilibrium vapor pressures of the liquid in the pores and that of the bulk liquid, respectively. The interaction between the pore walls and the adsorbate is based on physisorption (van der Waals interaction) and leads to the formation of adsorbate layers at low p/p_0. Usually, the amount adsorbed on the porous solid under study is plotted as a function of the amount adsorbed on an ideal nonporous reference solid with similar surface characteristics, providing parameters such as the overall pore volume, specific surface area and micropore volumes. Gas adsorption method may be carried out using a system similar to the one for the BET method as shown in Figure 3.22.

3.2.2.8 *Positron annihilation*

Positron annihilation lifetime spectroscopy (PALS) is a powerful tool for the detection and quantification of defects on the atomic scale in various types of solids. PALS is sensitive to different kinds of defects, such as dislocations and vacancies in metals or crystals, grain boundaries, voids and pores. Similar to scattering techniques, PALS is a noninvasive technique and allows the detection of inaccessible pores. As shown in Figure 3.24, PALS is based on positron annihilation, the decay of positrons into two γ photons. With ^{22}Na as the radioactive source, the formation of positrons (β^+) by radioactive decay is accompanied by the simultaneous emergence of a γ-quantum of 1.273 MeV, which defines the starting signal of the positron lifetime measurement. If a sample is introduced, the positrons lose their high energy by inelastic collisions with electrons.

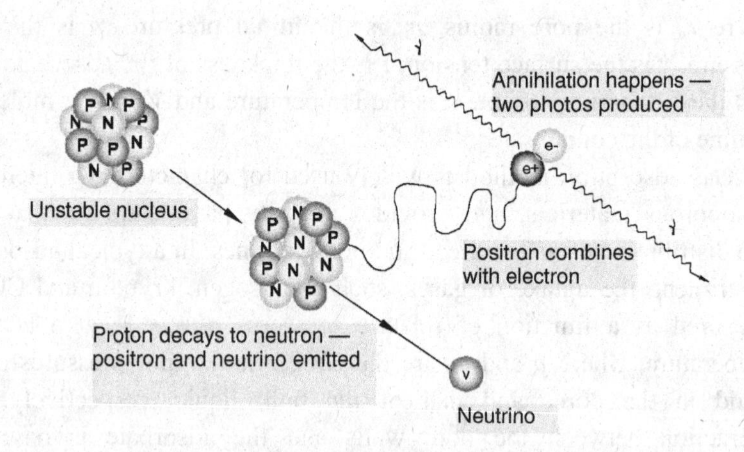

Figure 3.24 Principle of positron annihilation [32].

These "thermalized" positrons, with energies in the order of a few meV, form positroniums (Ps, the electron-positron bound state). The lifetime, the inverse of the annihilation rate, becomes longer when a positron or positronium is localized at a space with lower electron density such as a void. Therefore, positrons can be used as a probe to investigate the average sizes of the free volume, the size distribution and the free volume concentration by measuring their lifetimes.

3.2.2.9 *Mercury porosimetry*

In mercury porosimetry (MP), gas is evacuated from the sample, and the sample is then immersed in mercury. As mercury is a nonwetting fluid at room temperature for most porous materials of technological interest, an external pressure is applied to gradually force the nonwetting mercury into the sample. By monitoring the incremental volume of mercury intruded for each applied pressure, the pore size distribution of the sample can be estimated in terms of the volume of the pores intruded for a given radius r. The evaluation of pore sizes from MP is based on the Washburn equation:

$$p = \frac{2\sigma \cos\theta}{r} \tag{3.2}$$

where p is the pressure required to force a nonwetting fluid into a circular cross-sectional capillary of radius r, σ is the surface tension and θ is the wetting angle.

Mercury porosimetry allows the determination of average pore sizes in the range between 3 nm and 200 nm and their distributions. In addition, this method overestimates the volume of the smallest pores in the case of ink-bottle-shaped pores, because the intrusion of mercury into the larger pores is determined by the small openings. Moreover, it should be pointed out that pore size distributions by the Washburn equation are not a geometrical relationship, but a physical characteristic of a porous medium, because mercury porosimetry is based on transport and relaxation phenomena.

Figure 3.25 shows three main steps in mercury porosimetry. First, the sample is placed in the column which is evacuated subsequently. In next step mercury is filled. In the third step, the pressure is applied and values are taken to estimate the pore sizes.

3.2.3 *Nanoparticles in biological systems*

Most of the characterization methods discussed above are specifically meant for solid phase nanoparticles. However, on examination of biomolecules in living systems, it is found that the nanoparticles in liquid phase hold the key. If the biomolecules are examined from bottom-up

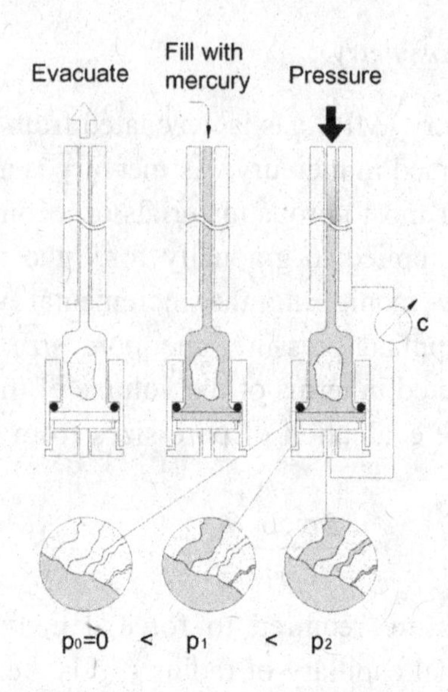

Figure 3.25 Principle of mercury porosimetry (IBP Holzkirchen, Germany).

approach of nanotechnology, it can be found that many biomolecules comprise of nanoliquids. Before nanoanalytical methods such as electron microscopy can be applied, biological samples often require complex sample preparation steps.

The transmission electron microscopy (TEM) is widely used in materials research, and is an important tool in analyzing biological macromolecules, such as proteins and viruses, on the molecular level. TEM also plays an important role in analyzing pharmaceutical, horticulture and agriculture products. There are two major problems when TEM is used for biological samples. One problem is that, since the path of the electron beam must be in vacuum, the biological sample to be observed must also be kept in vacuum. The effect of vacuum on biological specimens requires special analysis and study. The other problem is that the electron beam radiations may cause severe damage to the specimen under observation. These difficulties can be overcome by keeping the sample in the hydrated state by ice embedding method in which specimen is frozen. That is the reason why cryogenic transmission electron microscopy (Cryo-TEM) method is often used for investigating biomolecules.

Scanning transmission ion microscopy (STIM) is another important method for analyzing biological sample. This method was developed based on the idea of scanning transmission electron microscopy (STEM). In this spectroscopy, an ion beam is used instead of an electron beam which is used in STEM. The main advantage of STIM is its greater penetration depth that allows the analysis of much thicker objects. This technique makes imaging as well as mass normalization possible at resolution down to 100 nm.

Cryo-TEM and STIM are two very important methods for nanoparticle detection in biological samples, and they have been extensively used in biomedical research. For example, these methods have been applied to assess whether the nanoparticles used in cosmetics, such as creams and sunscreens, can penetrate into human skins and consequently cause systemic effects.

References:

1. Paulose, M., Varghese, O. K. and Grimes, C. A. (2003). Synthesis of gold-silica composite nanowires through solid-liquid-solid phase growth, *Journal of Nanoscience and Nanotechnology*, **3**, 341-346.
2. Chang, P. C., Fan, Z., Wang, D., Tseng, W. Y., Chiou, W. A., Hong, J. and Ju, J. G. (2004). ZnO nanowires synthesized by vapor trapping CVD method, *Chemistry Materials*, **16**, 5133-5137.
3. Park, J. Y., Lee, D. J. and Kim, S. S. (2005). Size control of ZnO nanorod arrays grown by metalorganic chemical vapour deposition, *Nanotechnology*, **16**, 2044-2047.
4. Yu, D. P., Lee, C. S., Bello, I., Sun, X. S., Tang, Y. H., Zhou, G. W., Bai, Z. G., Zhang, Z. and Feng, S. Q. (1998). Synthesis of nano-scale silicon wires by excimer laser ablation at high temperature, *Solid State Communications*, **105** (6), 403-407.
5. Cheng, X. and Guo, L. J. (2004). One-step lithography for various size patterns with a hybrid mask-mold, *Microelectronic Engineering*, **71**, 288-293.
6. Melville, D. O. S. and Blaikie, R. J. (2005). Super-resolution imaging through a planar silver layer, *Optics Express*, **13**, 2127-2134.
7. Chen F. T. (2003). Asymmetry and thickness effects in reflective EUV masks, *Proceedings of SPIE*, **5037**, 347-354.
8. Colburn, M. (2001). Development and advantages of step-and-flash imprint lithography, *Solid State Technology*, **46**(7), 67-78.
9. Choi, B. J. (2001). Distortion and overlay performance of UV step and repeat imprint lithography, *Proceedings of the Micro- and Nano-Engineering International Conference'* 2004, Rotterdam, Netherlands.

10. Cardinale, G.F. and Albert A.T. (2006). *Programmable imprint lithography template*, U. S. Patent 7,128,559.
11. Michel, B. Bernard, A., Bietsch, A., Delamarche, E., Geissler, M., Juncker, D., Kind, H., Renault, J. P., Rothuizen, H., Schmid, H., Schmidt-Winkel, P., Stutz, R. and Wolf, H. (2001). Printing meets lithography: Soft approaches to high-resolution, *IBM Journal of Research and Development*, **45** (5), 697-719.
12. Carter, D. J. D., Pepin, A., Schweizer, M. R. and Smith H. I. (1997). Direct measurement of the effect of substrate photoelectrons in X-ray nanolithography, *Journal of Vacuum Science and Technology B*, **15**, 2509-2513.
13. Vladimirsky, Y., Bourdillon, A. J., Vladimirsky, O., Jiang, W. and Leonard, Q. (1999). Demagnification in proximity X-ray lithography and extensibility to 25 nm by optimizing Fresnel diffraction, *Journal of Applied Physics D*, **32**, L114-L118.
14. Bourdillon, A. and Vladimirsky, Y. (2006). *X-ray Lithography on the Sweet Spot*, UHRL, San Jose.
15. Schattenburg, M. L., Early, K., Ku, Y. C., Chu, W., Shepard, M. I., THE, S. C., Smith, H. I., Peters, D. W., Frankel, R. D., Kelly, D. R. and Drumheller, J. P. (1990). Fabrication and testing of 0.1-mu-m-linewidth microgap X-ray masks, *Journal of Vacuum Science and Technology B*, **8** (6), 1604-1608.
16. Kruk, M., Jaroniec, M., Ryoo, R. and Kim, J. M. (1997). Monitoring of the structure of siliceous mesoporous molecular sieves tailored using different synthesis conditions, *Microporous Materials*, **12**, 93-106.
17. Gidley, D. W., Frieze, W. E., Dull, T. L., Yee, A. F., Ryan, E. T., and Ho, H. M. (1999). Positronium annihilation in mesoporous thin films, *Physical Review B*, **60**, R5157-R5160.
18. Bragg, W. H. and Bragg W. L. (1913). The reflection of X-rays by crystals, *Proceedings of the Royal Society of London. Series A*, **88**, 428-438.
19. Seiler, H. (1983). Secondary-electron emission in the scanning electron microscope, *Journal of Applied Physics*, **54**, R1-R8.
20. Brunauer, S., Emmet, P. H. and Teller, E. (1938). Adsorption of gases in Multimolecular Layers, *Journal of the American Chemical Society*, **60**, 309-319.
21. Wu, M. C., Aziz, A., Witt, J. D. S., Hickey, M. C., Ali, M., Marrows, C. H., Hickey, B. J. and Blamire, M. G. (2008). Structural and functional analysis of nanopillar spin electronic devices fabricated by 3D focused ion beam lithography, *Nanotechnology*, **19**, 485305.
22. Bouchiat, V. and Esteve, D. (1996). Lift-off lithography using an atomic force microscope, *Applied Physics Letters*, **69** (20), 3098-3100.
23. Sweeney, D. (1999). Extreme ultraviolet lithography: imaging the future, *Science and Technology Review*, November, 4-9.
24. AMO GmbH, http://www.amo.de/imprint_process.0.html?&L=1.
25. www.photon-production.co.jp/e/ppl-mirrircle-20sx.html.
26. http://jleenano.snu.ac.kr/final/?sid=312.
27. www.piaggio.ccii.unipi.it//bio/biochem/index.php?id=soft%20Lithography.htm.

28. Plusquellic, D. F. (2006). High-resolution UV spectroscopy, http://physics.nist.gov/Divisions/Div844/facilities/uvs/uvs.html.
29. http://midas.npl.co.uk/midas/content/mn027.html.
30. www.rpi.edu/dept/materials/COURSES/NANO/shaw/page5.html.
31. Jacobs, R. Basic operating principles of the Sorptomatic 1990, http://saf.chem.ox.ac.uk/Instruments/BET/sorpoptprin.html.
32. Badawi, R. (1999). *Introduction to PET Physics*, http://depts.washington.edu/nucmed/IRL/pet_intro/.

Chapter 4

Carbon Nanomaterials

Carbon nanomaterials are among the most extensively investigated and applied nanomaterials [1]. In this chapter, after introducing the typical carbon allotropes, the synthesis, properties and applications of three types of carbon nanomaterials will be discussed: fullerences, carbon nanotubes and carbon nanofoams.

4.1 Carbon Allotropes

Carbon is one of the most versatile elements with a wide variety of stable forms, possessing unique properties and many wonderful applications. The unique character of carbon stems from its ability to form distinct types of valance bonds among themselves, leading to different physical properties. As shown in Figure 4.1, pure carbon has four typical crystalline forms: diamond, graphite, fullerenes and nanotubes [2]. Other common carbon includes amorphous carbon, charcoals, soot and glassy carbon which are microcrystalline forms of graphite. There are still other special carbon nanostructures, such as nanoballs and nanofibers, which can be applied to special fields.

Graphite and diamond are two main allotropic forms of carbon. Graphite consists of single atom layers of carbon with sp^2 hybridization in a honeycomb pattern. This state of solid carbon is thermodynamically stable, exhibiting semimetallic behaviors. Diamond has a tetrahedral structure with sp^3 bonded carbon atoms, so it is an isotropic cubic wide gap insulator. Diamond-like carbon is a metastable and amorphous form of carbon containing a mixture of both sp^2 and sp^3 bonding and is technologically important as hard, chemically inert, and insulating coatings.

Fullerenes are a new form of carbon invented by Kroto *et al.* in 1985 [3]. A fullerene is a single molecule consisting of 60 carbon atoms (C_{60}) which are arranged in the shape of a soccer ball. Although carbon nanotubes (CNTs) are proper fullerenes, their discovery did not stem directly from the discovery of C_{60}. Instead, nanotubes have their roots in the pyrolytic and vapor phase deposition processes in which conventional carbon fibers are historically grown. CNTs was discovered by Sumio Iijima, a Japanese electron microscopist working at the NEC Corporation, in 1991 [4]. He provided a rigorous structural solution to fibers created by arc discharge, and described "helical microtubules of graphite carbon having outer diameters of 4-30 nm and lengths of up to 1 microns".

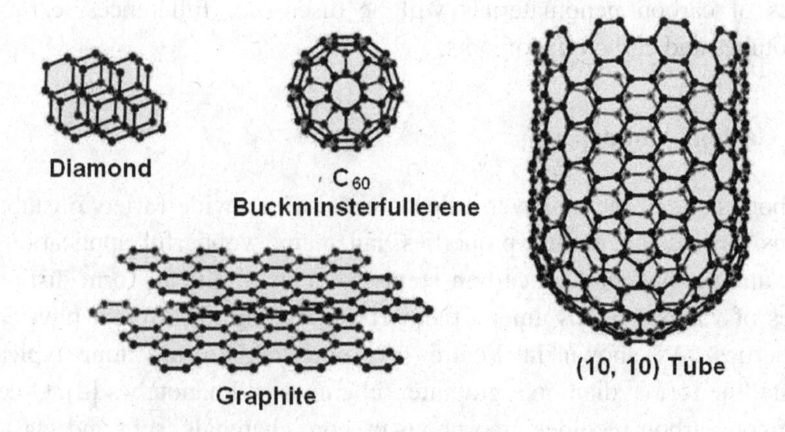

Figure 4.1 Four typical allotropes of carbon: graphite, diamond, fullerene and nanotube [2].

Besides the four typical allotropes of carbon discussed above, carbon nanofoam is the fifth allotrope of carbon. Carbon nanofoam was discovered by Andrei V. Rode and his team at the Austrian National University in Canberra [5, 6]. The molecular structure of carbon nanofoam consists of carbon tendrils bonded together in a low-density, mist-like arrangement. With a density of about 2 mg/cm^3, carbon nanofoam is one of the lightest known solid substances. It very interesting that carbon nanofoam exhibits unusual ferromagnetism.

4.2 Fullerenes

The discovery of fullerenes has revolutionized materials research and has given a new dimension to carbon and its compounds. Since the discovery of carbon nanotubes by Iijima, they are of great interest, both from a fundamental point of view and for future applications [7, 8]. The most important features of these structures are their electronic, mechanical, optical and chemical characteristics, which open a way for future applications. These properties can be measured on single nanotubes. However, for commercial application, large quantities of purified nanotubes are needed.

Fullerenes consist of a spherical, ellipsoid or cylindrical arrangement of dozens of carbon atoms. Fullerenes were named after Richard Buckminster Fuller, an architect known for the design of geodesic domes which resemble spherical fullerenes in appearance [9]. Spherical fullerenes are often called "buckyballs", whereas cylindrical fullerenes are known as "buckytubes" or "nanotubes". Usually the term "fullerene" refers to molecules such as C_{28}, C_{32}, C_{44}, C_{50}, C_{58}, C_{60}, C_{70}, C_{72}, C_{78}, C_{80}, C_{82} and so on, in which all the carbon atoms are on a spherical or spheroidal surface. In these molecules, carbon atoms are located on the vertices of exact hexagons or pentagons that coat an orbital or spheroid surface. The major interest in fullerene molecules is generated because they have a series of unusual properties. Fullerenes are similar in structure to graphite, which is composed of a sheet of linked hexagonal rings, but they contain pentagonal (or sometimes heptagonal) rings that prevent the sheet from being planar. Fullerenes occur only in small amounts naturally, but several techniques for producing them in greater volumes have been developed. In the early gas phase work, the fullerenes molecules were produced by the laser vaporization of carbon from a graphite target in a pulsed jet of helium.

4.2.1 *Molecule structures*

According to the isolated pentagon rule, fullerenes are grown by partition of pentagons by hexagons. The C_{60} has a structure of a truncated icosahedron, like a soccer ball made out of hexagons and pentagons with

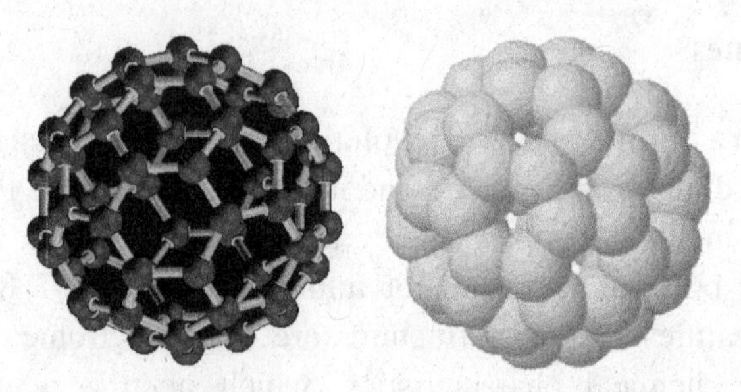

Figure 4.2 The structure of fullerene C_{60} [73].

carbon atoms at the corners of each hexagon and a bond along each edge. Figure 4.2 schematically shows the placement and position of pentagons and hexagons in fullerene molecule C_{60} [10]. In a C_{60} molecule, each carbon atom is triagonally bonded to three other carbon atoms. Among the 32 faces on the regular truncated icosahedron, 20 are hexagons, and the remaining 12 are pentagons. Therefore, a C_{60} molecule can be considered as a "rolled-up" graphite sheet which forms a closed shell. This is in close agreement with Euler's theorem, which states that a closed surface consisting of hexagons and pentagons has exactly 12 pentagons and an arbitrary number of hexagons.

The introduction of pentagons gives rise to curvature in forming a closed surface. To minimize local curvature, the pentagons become separated from each other in a self-assembly process. This gives rise to the isolated pentagon rule. In terms of mathematics, fullerenes have a trivalent convex polyhedron structure with pentagonal and hexagonal faces. According to Euler's theorem, we have

$$F - E + V = 2 \tag{4.1}$$

where F, E and V are numbers of faces, edges and vertex respectively. Therefore, the number of pentagons in C_{60} can be arrived at by the fact that every vertex in fullerene structure is made out of three faces. C_{60} contains 20 hexagons. C_{60} molecules obey the isolated pentagon rule: none of the pentagons make contact with each other.

The number of fullerenes C_{2n} grows rapidly with increasing value of n. The smallest possible fullerene is the dodecahedral C_{20} [11]. As shown in Figure 4.3, a C_{20} molecule consists of 12 pentagonal faces and no hexagonal face.

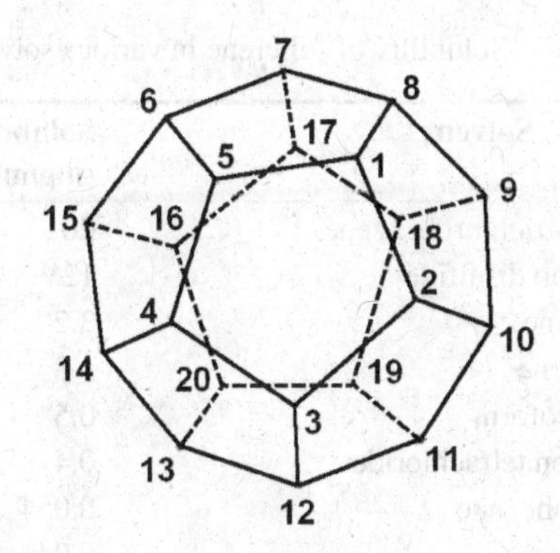

Figure 4.3 The structure of fullerene C_{20}.

4.2.2 *Physical and chemical properties*

In the study of the physical and chemical properties of fullerenes, it is important to note that all the 60 carbon atoms in C_{60} are chemically equivalent. The X-ray diffraction (XRD) analysis of C_{60} indicates that the center to center distance between adjacent C_{60} molecules is 10 Å and the molecular diameter is 7.1 Å which is consistent for the van der Waals interatomic distance of 2.9 Å.

Fullerenes are conductor and heat resistant. The first reported superconductor in the fullerene family is $K_3 C_{60}$ at the temperature of 19 K, and subsequent work has revealed superconductivity at the temperature of 40 K in $Cs_3 C_{60}$ under a pressure of 12 kbar [12, 13]. The unusual structure of molecular fullerenes leads to unusual optical transport and magnetic properties, with possible applications such as photoconductors, diode rectifiers, optical limiters, photorefractive materials, bonding agents, STM tip coatings.

Due to the graphite-like bonds, fullerenes are not very reactive, and are sparingly soluble in many solvents [14, 15]. At room temperature, fullerenes are the only carbon allotrope that can be dissolved in common solvents, such as benzene, toluene, xylene and carbon disulfide. Solutions of pure C_{60} fullerenes have a deep purple color, while those of C_{70} are reddish brown. The higher fullerenes C_{76} to C_{84} have a variety of colors. As some fullerenes, such as C_{36} and C_{50}, have a small band gap between the ground and excited states, they are not soluble. Table 4.1

Table 4.1 Solubility of fullerene in various solvents.

Solvent	Solubility (mg/ml)
1,2,4-trichlorobenzene	20
Carbon disulfide	12
Toluene	3.2
Benzene	1.8
Chloroform	0.5
Carbon tetrachloride	0.4
Cyclohexane	0.054
N-hexane	0.046
Tetrahydrofuran	0.037
Acetonitrile	0.02
Methanol	0.0009

lists the solvents that can dissolve a fullerene extract mixture (C_{60}/C_{70}). The solubility is represented as approximate saturation of fullerene (mg/ml) in solvent.

A spherical fullerene of n carbon atoms has n pi-bonding electrons. These should try to delocalize over the whole molecule. The quantum mechanics of such an arrangement should be like one shell of the well-known quantum mechanical structure of a single atom, with a stable filled shell for n = 2, 8, 18, 32, 50, 98, 128, etc, i.e. twice a perfect square; but this series does not include 60. As a result, C_{60} in water tends to pick up two more electrons and become an anion.

Fullerenes tend to react as electrophiles. An additional driving force is the relief of strain when double bonds become saturated. The key in this type of reaction is the level of functionalization i.e. mono addition or multiple additions. In case of multiple additions their topological relationships huddled together or evenly spaced. In some cases, the opening of fullerenes by breaking several of the double bonds takes place, so that small molecules can be inserted through the hole, for instance, hydrogen in endohedral hydrogen fullerene.

Fullerenes react as electrophiles with a host of nucleophiles in nucleophilic additions. Examples of nucleophiles are Grignard reagents and organolithium reagents. The reaction of C_{60} with methylmagnesium chloride stops quantitatively at the penta-adduct with the methyl groups centered around a cyclopentadienyl anion which is subsequently protonated. Another nucleophilic reaction is the Bingel reaction. The 6-6 bonds of fullerenes react as dienes or dienophiles in cycloadditions, for instance, Diels-Alder reactions.

Fullerenes are resistant to hydrogenation. Though it is more difficult than reduction, oxidation of fullerenes is possible, for instance, with oxygen and osmium tetraoxide. Fullerenes also react in electrophilic additions. The reaction with bromine can add up to 24 bromine atoms to the sphere. Fullerenes react with carbenes to methanofullerenes as well. Fullerenes are a ligand in organometallic chemistry. The 6-6 double bond is electron-deficient and forms metallic bonds. In direct sunlight, C_{60} fullerenes react with tungsten hexacarbonyl $W(CO)_6$ and form complexes in hexane solutions.

4.2.3 *Synthesis methods*

Many methods have been developed for the synthesis of fullerences. In the following, we discuss three typical methods: electric arc method, laser ablation method and solar energy method.

4.2.3.1 *Electric arc method*

Fullerenes can be synthesized using the arc discharge between graphite electrodes (20 V, 60 A) in approximately 200 Torr of He gas. The heat generated at the contact point between the electrodes causes evaporation of carbon to form soot and fullerenes, which condense on the water-cooled walls of the reactor. This discharge produces a carbon soot which can contain up to 15% fullerenes: C_{60} (13%) and C_{70} (2%). As the mass of a fullerence molecule is proportional to the number of carbon atoms in the molecule, the fullerenes synthesized can then be separated from the soot according to their mass using liquid chromatography.

4.2.3.2 *Laser ablation method*

This method uses a pulsed high-energy laser beam to evaporate a graphite target, and in this way, the percentage of the fullerenes which are photochemically destructed will be decreased. As shown in Figure 4.4, a laser beam, focused at 45° to the target surface, is used to ablate the target. This method is considered to be the most efficient for producing fullerenes [16].

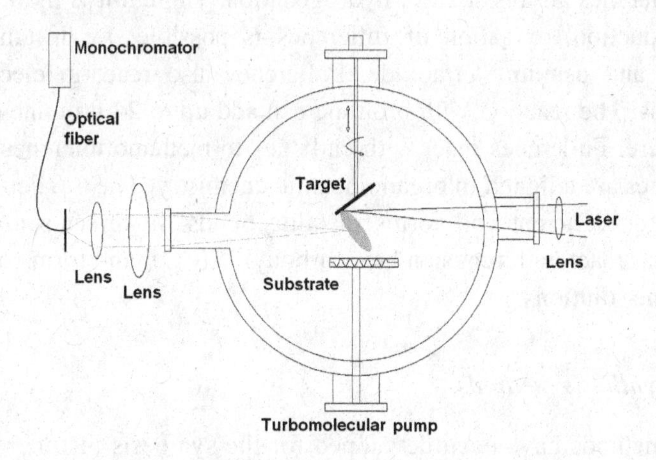

Figure 4.4 Schematic of laser ablation process for fullerene synthesis [74].

4.2.3.3 *Solar energy method*

This method is similar to the electric arc method, except that the heating source for this method is the solar energy [17]. A ratio between the mass of fullerenes and the total mass of carbon soot defines the fullerene yield. The yields are determined by UV absorption techniques, and they are approximately 40%, 10-15% and 15% in laser, electric arc and solar processes, respectively. Interestingly, laser ablation technique has both the highest yield and the low productivity and, therefore, a scale-up to a higher power is costly. Actually, commercial production of fullerenes is a challenging task.

4.2.4 *Functionalization*

Functionalization of fullerenes has direct impact on their properties. For instance, a fullerene polymer can be obtained by adding a suitable polymerizable group, and the reactivity of fullerenes can be increased by attaching active groups to their surfaces. For understanding the functionalization of fullerenes, it is important to model the fullerenes. For the modeling, a new molecular dynamics method can be used. This model takes account of the charges at bonds and the electronic and atomic degrees of freedom. Functionalized fullerenes fall into two major classes: endohedral fullerenes with trapped molecules inside the cage and exohedral fullerenes with substituents outside the cage.

Ordered metal-coated fullerene clusters can be synthesized using a vapor synthesis method. For alkali metal-coated fullerenes, only monolayer coatings of metal atoms have been demonstrated, such as in the case of $Li_{12}C_{60}$ where lithium is believed to reside over the centers of each pentagonal face of C_{60}. If atoms of group III or rare earth are trapped inside the fullerenes, such structures are called endohedral fullerenes or endofullerenes. Figure 4.5 schematically represents the formation of endofullerene.

If a metal encapsulates the fullerenes, such structures are called endohedral metallofullerenes [18-20]. Metallofullerenes can be prepared by endohedral doping of guest species such as rare earth, alkaline earth or alkali metal ions into the interior of the fullerene molecule. The synthesis is carried out by impregnating the positive electrode with graphite powder mixed with the desired dopant. The trapping of foreign molecules causes the changes of electrical properties.

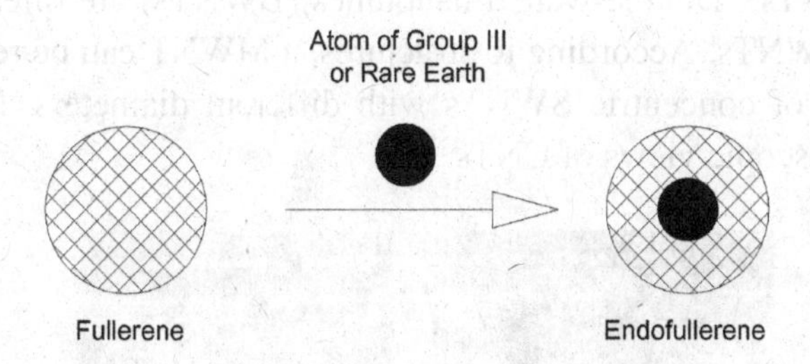

Figure 4.5 Formation of endofullerene.

4.2.5 *Possible risks*

Although fullerenes are relatively inert, it has been reported that they are injurious to organisms. For example, by producing free radicals in water that can damage the lipids on cellular membranes of animals, fullerenes can destroy the cells. There were also inflammatory changes in the liver and activation of genes related to the making of repair enzymes.

As many products in the future may contain fullerenes, methods for reducing toxicity of fullerenes should be investigated. It has been discovered that the toxicity can be reduced by an order of magnitude by adding hydroxyls, among other chemical groups.

4.3 Carbon Nanotubes

Carbon nanotubes (CNTs) are of great interest for both fundamental research and engineering applications [21]. They exhibit unique chemical, mechanical, electronic and optical characteristics, and these properties can even be measured on single nanotubes. Though carbon nanotubes exhibit great promise for future applications, large quantities of purified carbon nanotubes are needed for commercial applications.

4.3.1 *Why carbon nanotubes?*

Carbon nanotubes are tubular structures formed by special arrangement of carbon atoms. Generally speaking, there are two types of carbon nanotubes: single-walled nanotubes (SWNTs) and multi-walled nanotubes (MWNTs). Double-walled nanotubes (DWNTs) are often taken as a kind of MWNTs. According to structures, a MWNT can be regarded as a collection of concentric SWNTs with different diameters. Figure 4.6 shows microscopic views of CNTs.

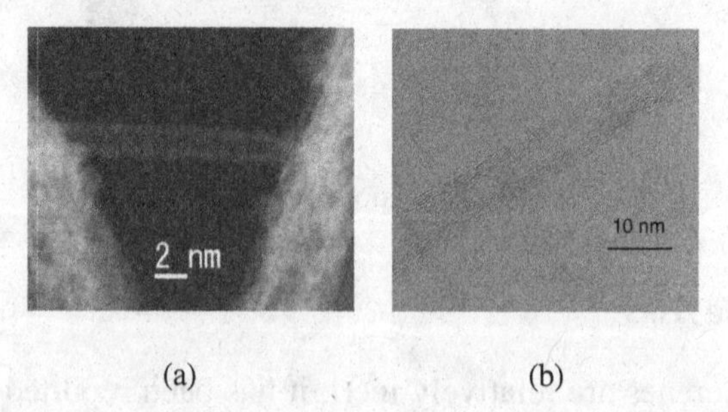

(a) (b)

Figure 4.6 TEM images of carbon nanotubes. (a) SWNT [71]. (b) MWNT.

Some of the key characteristics of CNTs include: high chemical and thermal stability, high conductivity, high capacitance and fast electron-transfer kinetics, high electrochemical sensitivity, strong electrocatalytic activity and highly efficient immobilization of biomolecules.

Carbon nanotubes are possibly the strongest material that can be made with known matter. They are about 100 to 150 times as strong as a steel with less than one fourth of the weight of a steel. This is a dramatic improvement over the carbon fibers used in developing the highest performance composites. Using nanotube epoxy composites, the weight of an airplane could be about one fifth of its present weight.

4.3.2 *Structure of carbon nanotubes*

Carbon nanotubes are often taken as quasi one-dimensional structures, as they usually have a length to diameter ratio of about 1,000. Usually, SWNTs have a diameter of 1 to 2 nm, whereas, MWNTs have an outer diameter of 10 nm or more, and an inner diameter of 1-3 nm. The length of SWNTs is usually less than 100 μm.

The appearance of single-walled nanotubes (SWNTs) is the true genesis of nanotechnology. The discovery that graphite can be rolled into a cylinder with a diameter of about one nanometer already has far-reaching consequences. Generally speaking, a SWNT consists of two separate parts with different physical and chemical properties: sidewall and end cap. The end cap structure is similar to a small fullerene, such as C_{60}. According to Euler's theorem, to obtain a closed cage structure consisting of only pentagons and hexagons, twelve pentagons are needed. The combination of a pentagon and five surrounding hexagons results in the desired curvature of the surface to enclose a volume. To have a stable structure, the distance between pentagons on the fullerene shell is maximized to obtain a minimal local curvature and surface stress. The sidewall of a SWNT is a cylinder, which is generated by wrapping a graphite sheet of a certain size in a certain direction [22].

Figure 4.7 shows SWNTs with different chiralities. The nomenclature (n, m) used to identify each SWNT refers to integer indices of two graphite unit lattice vectors corresponding to a nanotube's wrapping index, known as the chiral vector. As the dangling bonds at the edges of graphite sheets correspond to very high energy states, rolling of these sheets to cylindrical structures can cause minimization of total energy. Chiral vectors determine the directions along which the graphite sheets are rolled to form shell structures and are perpendicular to the tube axis vectors. The conductivity of a semiconducting nanotube changes as the nanotube is exposed to a miniscule amount of certain gas molecules. Gas molecules induce a charge transfer, which causes doping effects on semiconducting nanotubes. The gas molecules are adsorbed into the surface of the nanotube and each molecule induces small amounts (about 0.1 e) of electron transfer so that the nanotube becomes a p-type doped semiconductor.

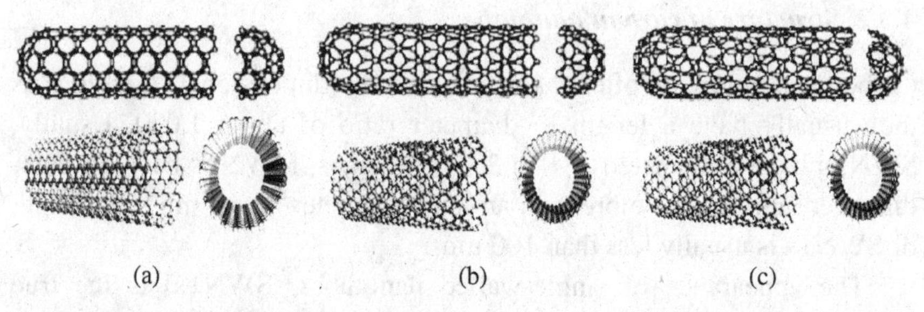

Figure 4.7 SWNTs with different chiralities [2]. (a) Armchair, (b) zigzag, (c) chiral structure.

Multi-walled nanotubes (MWNTs) are more common than single-walled nanotubes. As shown in Figure 4.8, an MWNT consists of concentric cylinders of graphite sheets, and its ends are also capped by half-fullerenes. Usually, the interlayer spacing between the concentric cylinders is about 0.3-0.4 nm, and the diameter of an MWNT is typically in the order of 10-40 nm. However, the lengths of these CNTs can be hundreds of microns to even millimeters.

4.3.3 *Special properties of carbon nanotubes*

Mainly due to their quasi-one dimensional structure, carbon nanotubes exhibit special properties [23-26]. In the following, we discuss the most important properties of CNTs in relation to their molecular structures.

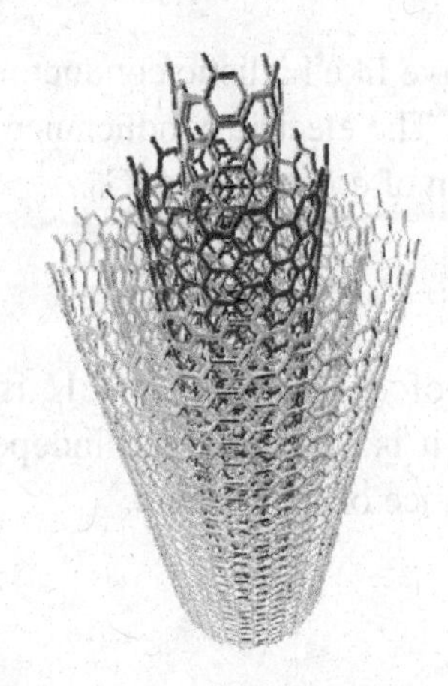

Figure 4.8 Rolled-up stack of graphite sheets for MWNT [21].

4.3.3.1 *Chemical reactivity*

Due to the curvature of its surface, a CNT is more chemically reactive than a graphite sheet. The reactivity of a CNT is closely related to the pi-orbital mismatch due to an increased curvature, and usually, a CNT with smaller diameter has higher reactivity. CNTs respond well to strong acids and other oxidizers which are believed to attach function groups such as -COOH, -OH, etc. to the side walls of nanotubes. Direct investigation of chemical modifications on nanotube behaviors will only be possible if they are pure, and the present difficulty is that the nanotubes synthesized in current research are not pure enough.

4.3.3.2 *Electrical conductivity*

A carbon nanotube with a small diameter can be metallic or semi-conducting. The difference in conductivity of carbon nanotubes are mainly due to their different chiral vectors which results in different band structures. The band structure of a CNT can be elucidated by a quantum mechanical derivation that begins with p_z orbital on a planar graphite sheet. It is found that, for a CNT, the transition from resistance to conduction is independent of its length.

A perfect metallic nanotube acts like ballistic conductor: when an electron is injected into one end, it comes out from other end of the nanotube. MWNTs behave like ballistic conductor despite the interaction between adjacent layers. The electric conductance is expected to be twice the fundamental quantum of conductance, G_0.

$$G_0 = \frac{2e^2}{H} \tag{4.2}$$

where e is the charge of an electron, and H is the Plank's constant. According to Eq. (4.2), it is expected that, independent of their lengths, MWCNTs have a resistance of 6,500 Ohm.

4.3.3.3 *Optical activity*

A defect-free CNT behaves like an optical fiber. A SWNT behaves like a single mode fiber for electrons. Theoretical studies reveal that, when a chiral CNT becomes larger, its optical activity may disappear, and its other physical properties may be also affected.

4.3.3.4 *Vibrational properties*

CNTs have symmetrical structures with many atoms. This cause large number of vibrational degrees of freedom and thus large number of phonon modes. At high symmetry, SWNT has a unit cell containing 160 atoms and results in 120 vibrational modes.

4.3.3.5 *Mechanical strength*

The Young's modulus of a carbon nanotube is related to its diameter and chirality. Generally speaking, carbon nanotubes have very large Young's modulus along their axial directions. The elastic moduli of CNTs are in the order of 1,000 GPa. Due to its small length scale, it is difficult to directly determine the Young's modulus of a CNT. Usually, the Young's modulus of a CNT is inferred from transmission electron microscopy.

CNTs are mechanically flexible. High bending angles, kinks and other deformation are possible in CNTs and these are fully reversible. CNTs have remarkable ability to accommodate deformation due to their ability of sp^2 carbon to rehybridize when subjected to any distortions which are out of plane.

4.3.3.6 *Specific heat and thermal conductivity*

Both the electrons ($C_{p,\ el}$) and phonons ($C_{p,\ ph}$) contribute to the specific heat of a carbon nanotube (C_p):

$$C_p = C_{p,el} + C_{p,ph} \qquad (4.3)$$

It has been shown that the ratio of $C_{p,ph}/C_{p,el}$ for SWNTs is approximately 100. This indicates that the contribution of phonon dominates and the specific heat (C_p) is approximately equal to specific contribution of phonon ($C_{p,ph}$). The inherent long range crystallinity of SWNTs results into better thermal conduction than MWNTs. It is difficult to predict specific heat of MWNTs by the above approach due to the complex structures of MWNTs.

The room temperature thermal conductivity of defect-free SWNTs is 6,000 W m^{-1} K^{-1}, while experimental data place this value in the range of 20-3,000 W m^{-1} K^{-1}.

4.3.4 *Synthesis of carbon nanotubes*

Many synthesis methods have been developed, such as chemical vapor deposition, vapor-liquid-solid process, sol-gel technique, electro-deposition method, hydrothermal/solvothermal processing, precipitation, template method, nanolithography, self-assembly technique, nano-imprinting, dip-pen lithography [21, 27-33]. However, their growth mechanisms of carbon nanotubes are yet not fully understood [34]. During the process of CNT formation, more than one mechanism may contribute together. In one of the mechanisms, metastable carbide particles (C_2), as a precursor, are formed on the surfaces of metal catalyst particles, and rod-like carbon structures are rapidly formed from the carbide particles.

In the following, we discuss four basic methods for synthesizing carbon nanotubes: arc discharge method, laser ablation method, chemical vapor deposition method and catalyzed decomposition method. Other methods can be taken as the permutations of these methods.

4.3.4.1 *Arc discharge method*

(1) *Carbon arc discharge method*

Though this method was first used for synthesizing C_{60} fullerenes, it may be the easiest way for the synthesis of carbon nanotubes [35]. But it produces a mixture of components, and a procedure for separating nanotubes from the soot is needed, and the catalytic metals exist in the rough products. In this method, carbon nanotubes are produced through arc-vaporization of two carbon rods placed end to end. Usually, the two carbon rods are separated by approximately 1 mm, and they are placed in an enclosure filled with inert gas, such as helium and argon, at pressure in the range of 50 to 700 mbar. Using this arc discharge method, nanotubes can also be synthesized in liquid nitrogen. Figure 4.9 shows a typical apparatus for synthesis of carbon nanotubes using carbon arc discharge method.

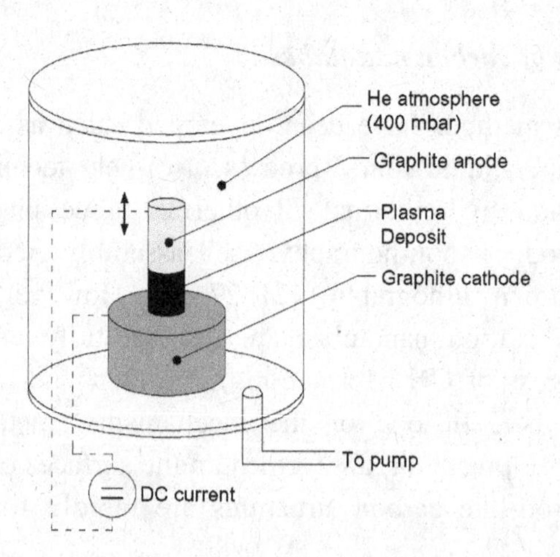

Figure 4.9 Apparatus for synthesizing carbon nanotubes using arc discharge method [21].

Depending on the electrodes used, we could selectively synthesize SWNTs or MWNTs. If both electrodes are graphite, the product is MWNTs, and if the anode is doped with Ni or Co or Fe, the product is SWNTs. The yield of carbon nanotubes is related to the uniformity of the plasma arc and the temperature of the deposit formed on the carbon electrodes.

(2) *Magnetic field synthesis*

Defect-free and high purity MWNTs can be synthesized in a magnetic field. In this way, the arc discharge synthesis is under control of a magnetic field around the arc plasma. In the magnetic field synthesis apparatus shown in Figure 4.10, the electromagnets are placed in the chamber of the apparatus to control the arc discharge [36]. Additional instrumentation, such as infrared (IR) spectrometer, may be attached to the chamber to collect the data and analyze the properties of the MWNTs obtained during the synthesis procedure. In this method, the electrodes used are graphite rods with extremely high purity (>99.999%), and MWNTs with purity higher than 95% can be achieved.

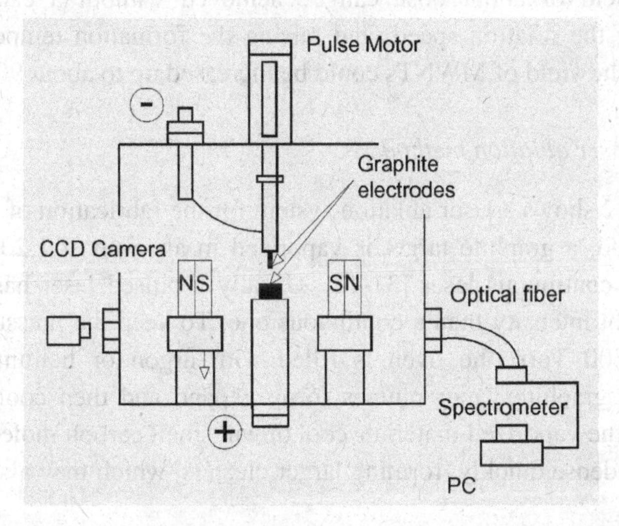

Figure 4.10 System for magnetic field synthesis of MWNTs [21].

(3) *Plasma rotating arc discharge method*

This method can be used for mass production of MWNTs. As shown in Figure 4.11, the rotation of the anode causes the centrifugal force, which generates turbulence and accelerates the carbon vapor perpendicular to

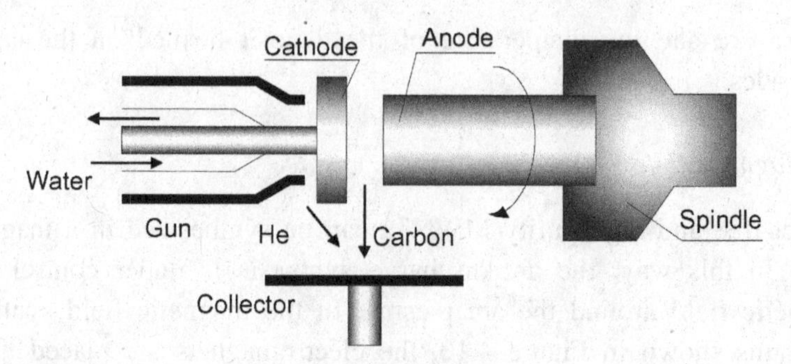

Figure 4.11 System for plasma rotating arc discharge [21].

the anode. Furthermore, due to the rotation of the anode, the micro discharges are uniformly distributed and stable plasma is generated, and thus the volume and temperature of the plasma are increased.

At a formation temperature of 1,025°C and a rotation speed of 5,000 rpm, a yield of about 60% can be achieved without a catalyst. By increasing the rotation speed, and raising the formation temperature to 1,150°C, the yield of MWNTs could be increased up to about 90%.

4.3.4.2 *Laser ablation method*

Figure 4.12 shows a laser ablation system for the fabrication of CNTs. In this method, a graphite target is vaporized in an oven at 1,200°C by a pulsed or continuous laser [37-42]. Usually a pulsed laser has a much higher light intensity than a continuous one. To keep the pressure of the oven at 500 Torr, the oven is filled with argon or helium. In this condition, graphite vapor plumes form, expand and then cool quickly. When all the vaporized materials cool down, small carbon molecules and atoms condense quickly, forming larger clusters, which may also include fullerenes.

In this method, the catalysts condense slowly and attach to carbon clusters, preventing them from forming cage structures. From these initial clusters, tubular molecules grow into single-walled carbon nanotubes until the catalyst particles become very large, or the temperature has reduced sufficiently so that carbon cannot diffuse over the surfaces of the catalyst particles. Due to the van der Waals forces,

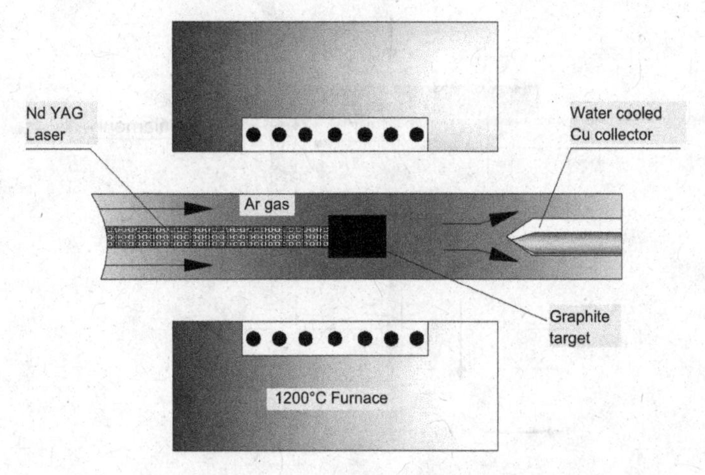

Figure 4.12 System for synthesis of CNTs by laser ablation [21].

the SWNTs formed in this way are usually bundled together, and contaminated with carbon nanoparticles.

4.3.4.3 *Chemical vapor deposition method*

In a chemical vapor deposition (CVD) method, a gaseous carbon source, such as acetylene, methane and carbon monoxide, is used [43-45]. An energy source, such as a heating coil, provides energy to the gaseous carbon molecules, cracking them into reactive atomic carbon. The carbon then diffuses towards the substrate, on which carbon nanotubes are formed [46]. Usually, the substrate is coated with a catalyst, such as Ni, Fe or Co, and in the synthesis procedure, it is heated to 650-900°C. Using this method, excellent alignment and positional control at the nanometer scale can be achieved. SWNTs can be synthesized by using an appropriate metal catalyst. This method has a yield of about 30%.

In the following, we will discuss five types of CVD often used in the fabrication of carbon nanotubes.

(1) *Plasma-enhanced chemical vapor deposition*

In a plasma-enhanced CVD method, a glow discharge is generated in a chamber by a high-frequency voltage, and during the discharge,

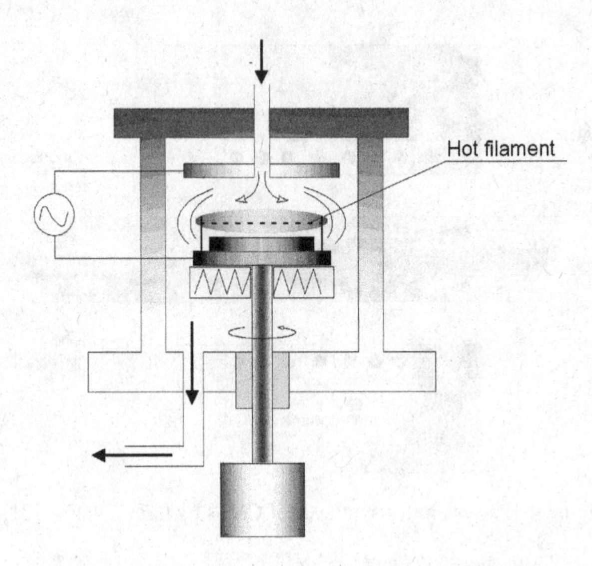

Figure 4.13 System for plasma enhanced CVD [21].

a carbon containing reaction gas, such as CO, C_2H_2, CH_4 and C_2H_4, is supplied to the chamber [47]. Figure 4.13 shows a typical plasma CVD system with a set of parallel plate electrodes. The reaction gas is supplied through the top electrode, and the substrate, such as Si, SiO_2 or glass, is put on the grounded electrode. The catalytic metal, such as Fe, Ni and Co are applied on the substrate using sputtering or thermal CVD. After nanoscopic metal particles are formed on the substrate, carbon nanotubes are grown on the metal particles by the discharge generated by the high frequency voltage.

The growth rate, morphology and microstructure of the carbon nanotubes are greatly affected by the catalyst used. It seems that nickel is the best pure-metal catalyst for the synthesis of aligned MWNTs [48]. The MWNTs synthesized in this method have a diameter of about 15 nm. The highest yield is about 50%, and it is achieved at temperatures below 330°C.

(2) *Thermal chemical vapor deposition*

In a thermal CVD method, the catalyst deposited on the substrate can be iron, nickel, cobalt or an alloy of these metals [49]. The catalyst deposited substrate is then etched by a diluted HF solution, and after that, the etched substrate is placed in a quartz boat which is positioned in a CVD reaction furnace, as shown in Figure 4.14. The catalytic metal film is further etched using ammonia gas at a temperature in the range of 750 to 1,050°C, and after that, nanometer-sized catalytic metal particles can be obtained, on which CNTs are grown.

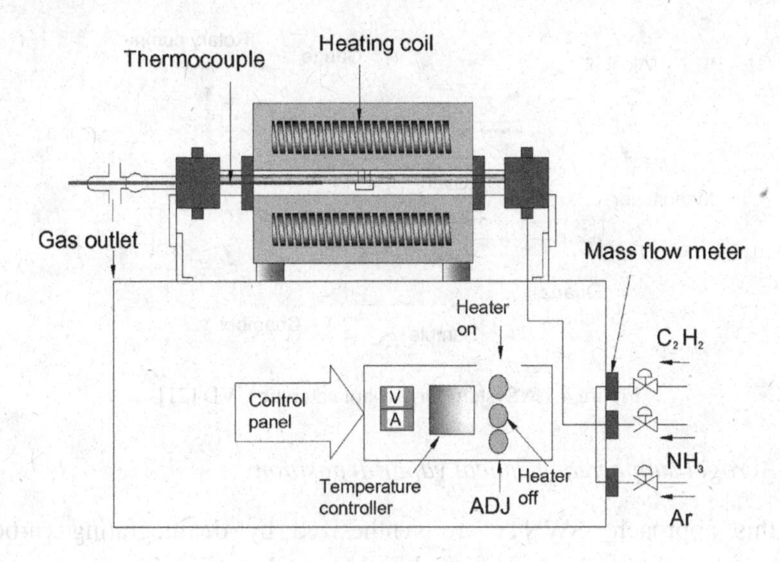

Figure 4.14 System for thermal CVD [21].

If iron catalyst is used, the diameter of the multi-walled carbon nanotubes synthesized by thermal CVD is related to the thickness of the catalytic film. If the catalytic film is 13 nm thick, the diameter of the nanotubes is in the range of 30 to 40 nm. If the catalytic film is 27 nm thick, the diameter of the nanotubes is in the range of 100 to 200 nm.

(3) *Alcohol catalytic chemical vapor deposition*

In this method, high quality SWNTs can be produced in large-scale and at low cost [50]. As shown in Figure 4.15, evaporated alcohol is applied over the iron and cobalt catalytic metal particles supported with zeolite. The reaction between the evaporated alcohol and the catalytic metal particles produces hydroxyl radicals, and the hydroxyl radicals can remove the carbon atoms with dangling bonds which are the obstacles in the synthesis of SWNTs. In this way, SWNTs can be synthesized at a temperature around 550°C, and the SWNTs fabricated have high-purity, and their diameter is about 1 nm. Furthermore, due to the low reaction temperature of this method, SWNTs can be directly grown on semiconductor devices that are already patterned with aluminum.

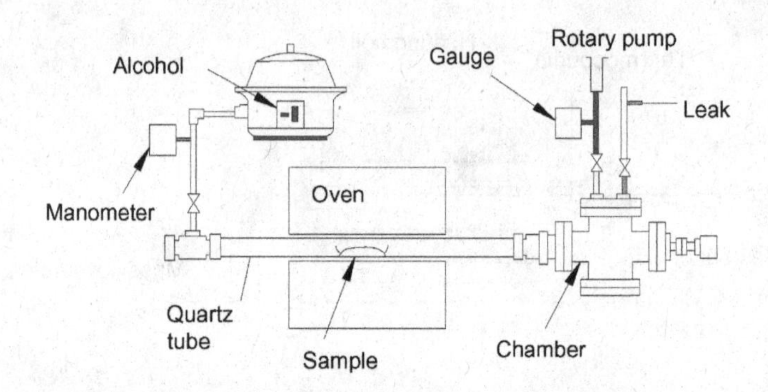

Figure 4.15 System for alcohol catalytic CVD [21].

(4) *Aerogel supported chemical vapor deposition*

In this approach, SWNTs are synthesized by disintegrating carbon monoxide on an aerogel supported Fe/Mo catalyst [51]. Many important factors, such as feeding gas, reaction temperature and the surface area of the supporting material, affect the yield and quality of SWNTs. As the aerogels have large surface area, high porosity and extremely low density, the catalyst has very high productivity. After an acidic treatment and oxidation process, SWNTs with purity higher than 99% can be obtained. The ideal reaction temperature for this method is about 860°C, and the nanotubes synthesized using this method have a diameter in the range of 1.0 to 1.5 nm.

(5) *Laser assisted thermal chemical vapor deposition*

In this method, a medium-power continuous wave CO_2 laser is used [52]. As shown in Figure 4.16, the laser beam, which is perpendicular to the substrate surface, pyrolyses the mixtures of

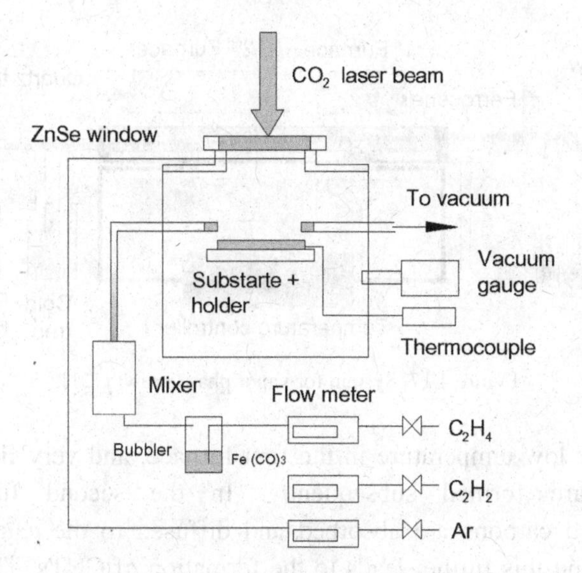

Figure 4.16 System for laser-assisted CVD [21].

$Fe(CO)_5$ vapor and acetylene in the flow reactor, and carbon nanotubes are formed under the catalyzing action of the small iron particles.

Usually the substrate is made of silica. Both SWNTs and MWNTs can be synthesized by using mixtures of iron pentacarbonyl vapor, ethylene and acetylene. The SWNTs synthesized in this method have a diameter in the range of 0.7 to 2.5 nm, and MWNTs synthesized in this method have a diameter in the range of 30 to 80 nm.

4.3.4.4 *Catalyzed decomposition method*

In a catalyzed decomposition method, organometallic catalyst precursor is fed along with the carbon source in gas phase. This method is different from CVD in such a way that CNTs do not grow on the catalyst.

(1) *Vapor phase growth*

In this method, the reaction gas and the catalytic metal are supplied directly to the chamber, and usually the catalyst used is ferrocene. In the vapor phase growth system schematically shown in Figure 4.17, there are two furnaces in the reaction chamber. The catalyst is vaporized at

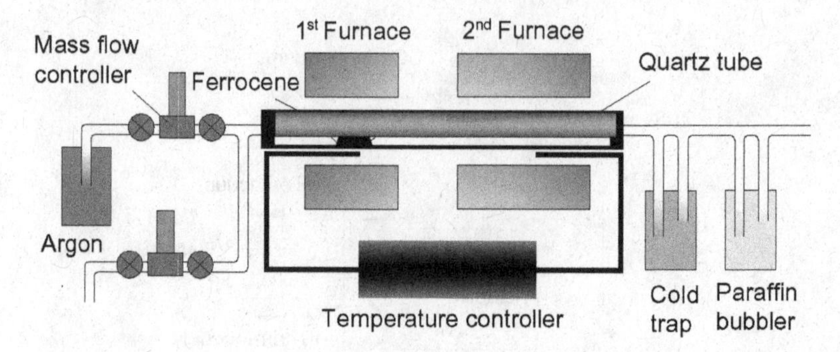

Figure 4.17 System for vapor phase growth [21].

a relatively low temperature in the first furnace, and very fine catalytic particles are formed subsequently. In the second furnace, the decomposed carbons are absorbed and diffused to the catalytic metal particles, and this further leads to the formation of CNTs. The diameter of the carbon nanotubes fabricated using this method is in the range of 2-4 nm for SWNTs and 70-100 nm for MWNTs.

(2) *CoMoCat method*

In this method, SWNTs are synthesized through Co disproportionation at temperatures in the range of 700-950°C. In this process, due to the formulation of a unique Co-Mo catalyst, the sintering of Co particles is inhibited, and thus the formation of undesired forms of carbon is restrained. Figure 4.18 shows a fluidized bed reactor which can be used for this method. An attractive advantage of fluidized bed reactors is that the continuous addition and removal of solid particles from the reactor is possible without stopping the operation. The fresh catalyst can be fed to the reactor column separately. The CO gas is fed from the bottom of the reactor column and the product is taken out from the top of the column. The residual CO gas is taken out from the top of the reactor column and recycled.

As CoMoCat catalyst has a high selectivity for SWNTs (80-90%), this method can be scaled-up without decreasing the quality of SWNTs. As listed in Table 4.2 [72], by changing the reaction conditions, such as temperature, SWNTs with different diameters can be obtained.

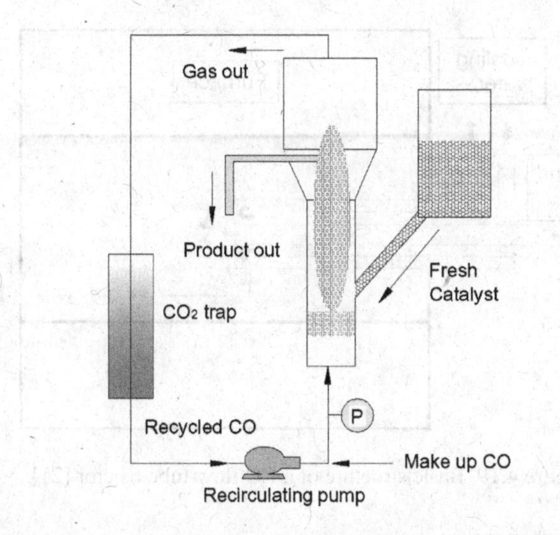

Figure 4.18 System for synthesizing carbon nanotubes using CoMoCat method [21].

Table 4.2 Diameter range versus temperature.

Temperature (°C)	Diameter range
750	0.9 ± 0.05
850	0.9-1.25
950	1.00-1.40

(3) *High pressure CO disproportionation process*

In this process, SWNTs are synthesized by flowing CO, as the carbon feedstock, mixed with a small amount of $Fe(CO)_5$, as the iron containing catalyst precursor, through a heated reactor. Figure 4.19 shows the layout of a CO flow-tube reactor. This process can be used for mass fabrication of carbon nanotubes.

The SWNTs fabricated using this method have an average diameter of about 1.1 nm. By tuning the pressure of CO, the size distribution of the carbon nanotubes can be controlled. Using this method, carbon nanotubes with diameter 0.7 nm, which are close to the smallest chemically stable SWNTs, have been produced. This method has a yield of about 70%, and SWNTs with purity about 97% can be synthesized at a rate of 450 mg/h using this technique.

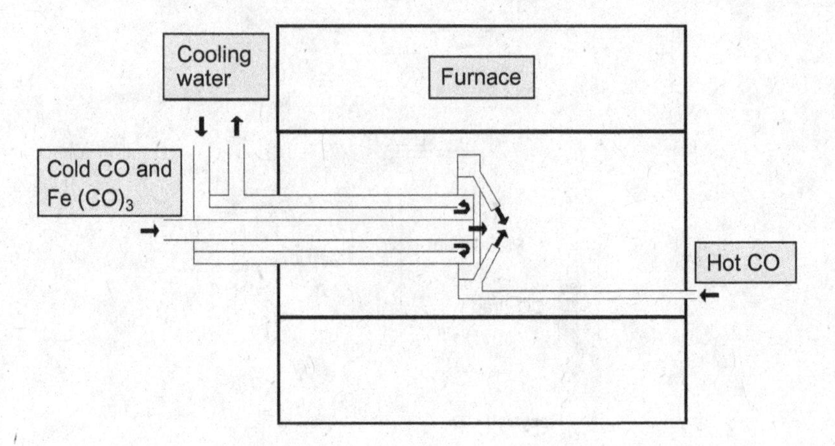

Figure 4.19 Basic structure of a CO flow-tube reactor [21].

(4) *Flame synthesis*

This method has its historic origin from India. The manufacturing of kohl is prevalent in India for many centuries. Kohl is a kind of black colored sticky material, called *Kajal*, often applied on the eye lids for cosmetic purpose as well as medicinal use. In this century-old process, an earthen dish is placed above the flame of vegetable oil lamp and kohl is collected on its surface. This kohl consists of carbon particles of various sizes including the nano scale.

However, for the production of SWNTs, this process has been developed and advanced with controlled flame environment. Under the controlled flame environment, suitable temperature can be obtained so that carbon atoms can be formed from the inexpensive hydrocarbon fuels and small aerosol metal catalyst islands can also be formed. In the same way as in the arc discharge method and the laser ablation method, SWNTs grow on the metal catalyst islands.

Generally speaking, there are three ways to make metal catalyst islands. The first way is to coat the metal catalyst islands, such as cobalt, on a mesh by physical vapor deposition. After exposure to a flame, these small island-like droplets become aerosols. In the second way, aerosol small metal particles are created by burning filter papers rinsed with a metal-ion solution, such as iron nitrate solution. The third way is the thermal evaporating technique. In this way, metal powders, such as iron or nickel, are inserted in a trough and heated.

4.3.5 *Purification of carbon nanotubes*

Because as-produced carbon nanotube soot contains a great deal of impurities, purification is very important for the applications of carbon nanotubes [53-59]. The impurities in carbon nanotube soot mainly include amorphous carbon, graphite sheets, smaller fullerenes and metal catalyst. These impurities interfere with the desired properties of SWNTs, and thus affects their applications. Besides, the fundamental research on SWNTs requires SWNTs be as pure as possible, and as homogeneous as possible. In the following, we discuss the methods often used for the purification of carbon nanotubes.

4.3.5.1 *Oxidation*

By oxidative treatment of SWNTs, carbonaceous impurities can be removed and metal surfaces can be cleared. One obvious disadvantage of this method is that the SWNTs can also be oxidized. Fortunately, compared to the damage to the impurities, the damage to SWNTs is very small.

Several techniques have been developed to protect SWNTs in the oxidation procedure. Using soluble oxidizing agents, such as H_2O_2 and H_2SO_4, mild oxidation in a wet environment will only oxidize the defects and will clear the metal surface. Another method is the oxidation by microwave heating. Microwave fields will heat up the metal, and the carbon attached to the metal will be catalytically oxidized.

4.3.5.2 *Acid treatment*

Generally speaking, the metal catalyst can be removed by acid treatment. In this method, the SWNTs to be treated are usually in a suspended form. Before the acid treatment, the surface of the metal should be exposed, for example, by sonication or oxidation. After that, the metal catalyst is exposed to acid and solvated. HNO_3 only has effects on the metal catalyst, and it will not damage SWNTs and other carbon particles. While HCl will have slight damage to SWNTs and other carbon particles.

4.3.5.3 *Annealing*

At high temperatures (600-1,600°C), the structures of carbon nanotubes will be rearranged, and the defects in carbon nanotubes will disappear. Furthermore, at high temperatures, the graphitic carbons and the short fullerenes will be pyrolysed [60]. By annealing carbon nanotubes at high temperature (1,600°C) and in vacuum condition, the metal catalysts will be melted and can be removed.

4.3.5.4 *Ultrasonication*

In this technique, different particles will become more dispersed due to ultrasonic vibrations [61]. The effects of ultrasonic vibrations are closely related to the surfactant, the solvent and the reagent used. The stability of the dispersed carbon nanotubes in a system is related to the solvent in the system. In a poor solvent, SWNTs attached to metal catalysts are more stable. While in some solvents, such as alcohols, monodispersed particles are more stable.

4.3.5.5 *Magnetic purification*

As most of the metal catalysts, such as iron, nickel and cobalt, are ferromagnetic, they can be mechanically removed from their graphitic shells by magnetic purification. In this method, inorganic nanoparticles, mainly ZrO_2 or $CaCO_3$, are mixed with the SWNT suspension in an ultrasonic bath to remove the ferromagnetic catalytic particles, which are subsequently trapped using permanent magnetic poles. SWNTs with high purity could be obtained after subsequent chemical treatments.

4.3.5.6 *Microfiltration*

Microfiltration is actually a kind of size separation. In this method, SWNTs are trapped by the filter, while other nanoparticles, such as catalyst metal, fullerenes and carbon nanoparticles, pass through the filter [62]. To separate SWNTs and fullerenes, the as-produced carbon soot is soaked in a CS_2 solution. As fullerenes are solvated in CS_2 solution, they pass through the filter. While as SWNTs are insoluble in CS_2 solution, they are trapped in the filter.

4.3.5.7 *Cutting*

SWNTs can be cut mechanically, chemically or by a combination of these two mechanisms [63]. By breaking the bonds between nanoparticles and nanotubes, ball milling can be used to mechanically cut nanotubes. In this way, the nanotubes will be disordered. SWNTs can also be cut in chemical way. In this method, nanotubes are partially functionalized, for example, with fluorene. The fluorated carbons are then driven off the sidewall with pyrolization in the form of CF_4 or COF_2, leaving behind the chemically cut nanotubes. CNTs can also be cut by ultrasonication in an acid bath, and this method is a combination of mechanical and chemical cuttings.

4.3.5.8 *Functionalization*

By attaching special functional groups to SWNTs, the nanotubes can be made more soluble than the impurities, such as metal catalysts. In this way, SWNTs can be easily separated from the impurities by filtration. The functional groups on SWNTs can be removed by thermal treatment, such as annealing, and the purified SWNTs can then be recovered. In this procedure, the SWNT structure is kept intact.

4.3.5.9 *Chromatography*

This method can be used to separate SWNTs into fractions with small length and diameter distribution, and the chromatographic techniques often used are Gel Permeation Chromatography (GPC) and High Performance Liquid Chromatography - Size Exclusion Chromatography (HPLC-SEC). In this method, SWNTs run over a column with a porous material, and the pore size will give a size distribution and size wise separation. SWNTs with larger sizes will come out earlier.

4.3.6 *Defects of carbon nanotubes*

As discussed above, carbon nanotubes can be fabricated and purified in various ways. However, economically feasible large-scale fabrication and purification techniques are still to be developed. The possible defects of carbon nanotubes should be taken into full consideration in the fabrication and purification procedures.

Besides the defects due to the impurities introduced during or after the nanotube growth process, deformations are another kind of defects, and seriously affect the properties of the nanotubes. Typical types of deformations, such as nanotube bends and nanotube junctions, are caused by replacing a hexagon with a heptagon or pentagon.

Defects may result in a variety of new structures, such as Y-branches, T-branches and junctions. Figure 4.20 schematically shows a Y-branch CNT. These defects can be formed in a controlled way for changing the electrical performances of carbon nanotubes.

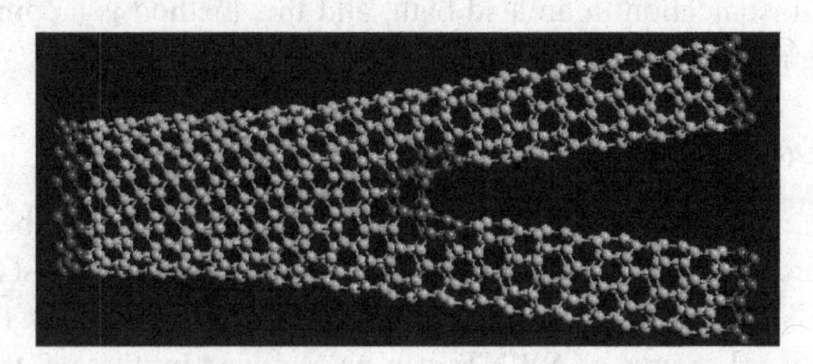

Figure 4.20 Schematic diagram of a Y-branch carbon nanotube [21].

4.4 Carbon Nanofoam

Carbon nanofoam is a low-density cluster-assembly of carbon atoms strung together in a loose three-dimensional web. Each cluster is about 6 nanometers wide, and is composed of about 4,000 carbon atoms linked in graphite-like sheets that are given negative curvature by the inclusion of heptagons among the regular hexagonal pattern. This is the opposite of the case of fullerenes. In the formation of fullerenes, the carbon sheets are given positive curvature by the inclusion of pentagons [5].

4.4.1 *Synthesis*

Carbon nanofoam can be fabricated using high-repetition-rate, high-power laser ablation of a glassy carbon sample in an argon atmosphere [6]. The process heats the sample to nearly 10,000°C, creating carbon vapor. When the vapor in the chamber reaches a threshold density, carbon clusters ranging from 6 to 9 nm in diameter begin to form near the hottest point and diffuse through the argon. The clusters connect randomly into a web-like foam that can then be collected from the walls of the chamber. Figure 4.21 shows the carbon nanofoam fabricated using this method.

Figure 4.21 Transmission electron micrograph (left), and scanning electron micrograph (right) of the carbon nanofoam, showing the web-like structure. The diffraction pattern in the inset shows the very broad rings indicating the lack of a long-range three-dimensional order in the foam. Reprinted with permission from Rode, A. V., Gamaly, E. G., Christy, A. G., Gerald, J. G. F., Hyde, S. T., Elliman, R. G., Luther-Davies, B., Veinger, A. I., Androulakis, J. and Giapintzakis, J. (2004). Unconventional magnetism in all-carbon nanofoam, *Physical Review B*, **70**, 054407. © 2004 by the American Physical Society.

4.4.2 *Properties*

Carbon nanofoam is practically transparent in appearance, consisting of mostly air, and fairly brittle. It has extremely high surface area and is capable of being exposed to thousands of Fahrenheit degrees before deforming. Carbon nanofoam is semiconducting with a band gap of 0.5 to 0.7 electron-volts.

One of the most unusual properties displayed by carbon nanofoam is that of ferromagnetism. This property fades within hours at room temperature; however, at temperatures about 90 K, the magnetism of carbon nanofoam may persists for up to 12 months and ranges from 0.36 to 0.8 electromagnetic units per gram. Figure 4.22 shows the magnetization curves of carbon nanofoam at different temperatures. Though the mechanism for the ferromagnetism of carbon nanofoam is yet to be fully understood, the magnetic properties of carbon nanofoam hints that the magnetism of a substance cannot be determined simply by the type of substance, but also by its allotrope and temperature [6].

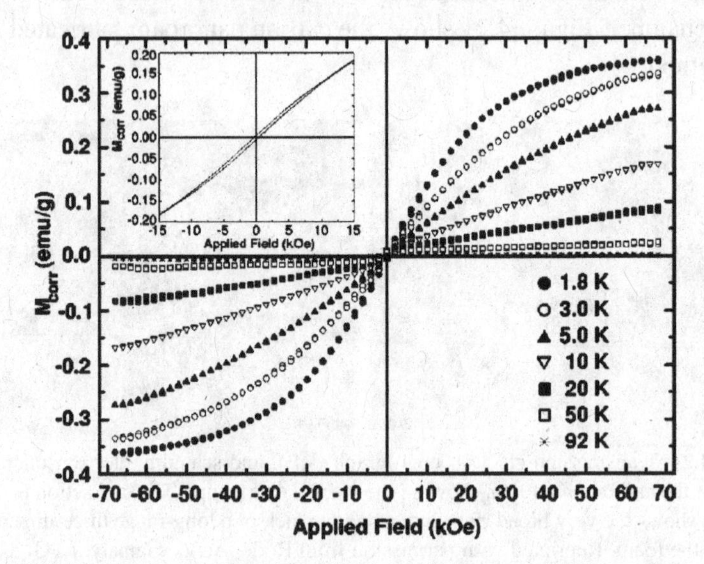

Figure 4.22 Mass magnetization of carbon nanofoam as a function of the applied magnetic field at several temperatures from 1.8 to 92 K. All data are corrected for the diamagnetic contribution of the gelatine sample holder. Inset: M(H) hysteresis loop at T = 1.8 K exhibiting a coercive force H_c = 420 Oe. Reprinted with permission from Rode, A. V., Gamaly, E. G., Christy, A. G., Gerald, J. G. F., Hyde, S. T., Elliman, R. G., Luther-Davies, B., Veinger, A. I., Androulakis, J. and Giapintzakis, J. (2004). Unconventional magnetism in all-carbon nanofoam, *Physical Review B*, **70**, 054407. © 2004 by the American Physical Society.

4.5 Applications

Carbon nanomaterials have exhibited great application potentials in various fields for different purposes, and many of them have been used for practical applications [64].

4.5.1 *Fullerenes*

Fullerenes have great potentials in medicinal applications [65-69]. For example, they can be bound with specific antibiotics for targeting certain cancer cells such as melanoma. The spherical shape of fullerenes, C_{60}, is very useful for molecular recognition. It can be used for inhibiting HIV proteases. Fullerenes and endofullerenes can efficiently photosensitize the conversion of oxygen into singlet oxygen in living cells, so they can be used as therapeutic agents as singlet oxygen can cleave DNA. The fullerenes can also be used as light-activated antimicrobial agents. Besides, gadolinium based endofullerenes are the potential MRI contrast agents.

4.5.2 *Carbon nanotubes*

Carbon nanotubes have been used in many fields. In the following, we discuss several typical applications examples of carbon nanotubes.

4.5.2.1 *Energy storage*

Carbonaceous materials have extensive electrochemical applications, such as batteries and fuel cells. Due to their small dimensions, smooth surface topology and perfect surface specificity, nanotubes are attractive advantages for energy storage. As nanotubes have the highest electron transfer rate, the fuel cells made of carbon nanotubes have very high efficiency.

4.5.2.2 *Hydrogen storage*

Hydrogen is a clean energy source. However, due to the volume and weight limitations, the storage of hydrogen is an important problem. Due to their cylindrical and hollow geometry, and nanometer-scale diameters, carbon nanotubes could store liquid hydrogen or hydrogen gas in the inner cores based on the capillary effect. The method of electrochemical storage can also be used for hydrogen storage. In this case, hydrogen atoms, instead of hydrogen molecules, are adsorbed, and this process is usually called chemisorption. Contrarily, the gas phase intercalation of hydrogen in CNTs involves the adsorption of H_2, which is usually called physisorption.

4.5.2.3 *Lithium intercalation*

The working principle of a rechargeable lithium battery is the electrochemical intercalation and deintercalation of lithium in both electrodes. An ideal battery has a high-energy capacity, fast charging time and a long cycle time. Lithium intercalated carbonaceous materials, such as graphite, are used in commercial Li-ion batteries. In these batteries, the specific energy capacity is partially limited by the thermodynamically determined equilibrium saturation composition of LiC_6. In the discharging (intercalation) procedure, following adsorption happens:

$$-C-+Li^{+}+e^{-}-C-Li_{ad} \tag{4.4}$$

Eq. (4.4) can be written in another way:

$$x\,Li+C_6 \rightarrow Li_x\,C_6 \tag{4.5}$$

In the charging (de-intercalation) procedure, the above process is reversed. Figure 4.23 schematically represents the adsorption in graphite.

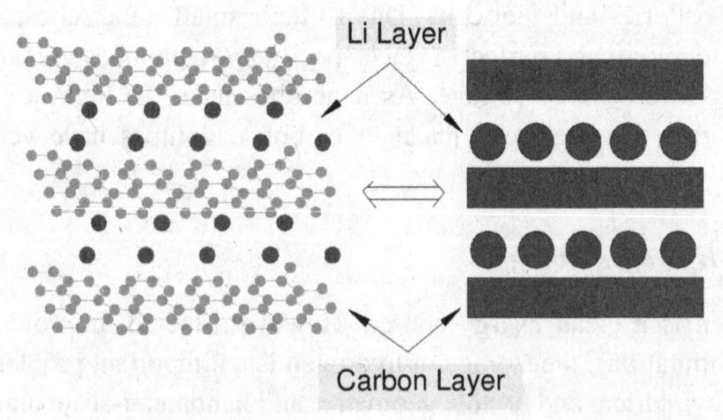

Figure 4.23 Structure of graphite intercalated with lithium [21].

Due to their special structure and chemical bonding, carbon nanotubes may be ideal intercalation hosts. Because guest species can intercalate in the interstitial sites and between the carbon nanotubes, carbon nanotubes are expected to have a higher saturation composition than graphite. Therefore, carbon nanotubes have the potential to be used as high energy density anode materials for rechargeable Li-ion batteries.

4.5.2.4 *Electrochemical supercapacitors*

Supercapacitors with extremely large capacitance are used in many electronic devices. Nanotubes are ideal for the development of supercapacitors because nanotubes have high surface areas accessible to electrolytes, and the electrode separation in the nanometer scale can be achieved.

4.5.2.5 *Molecular devices with CNTs*

Numerous molecular devices have been developed based on the special properties of carbon nanotubes. In the following, we discuss three typical types of molecular electronic devices with CNTs: field emitting devices, transistors and nanoprobes.

(1) *Field emitting devices*

If the electric field applied to a solid is sufficiently high, the electrons near the Fermi level can be extracted from the solid by tunneling through the surface potential barrier. The following characteristics make carbon nanotubes ideal for the development of field emitters: structural integrity, diameter in nanometer size, high electrical conductivity, small energy spread and high chemical stability. Field emitting devices based on carbon nanotubes have a lot of potential applications, such as flat panel displays, electron guns for electron microscopes and microwave amplifiers.

(2) *Transistors*

A single piece of semiconducting SWNT can be used to construct a field-effect transistor. If a suitable voltage is applied to the gate electrode, the nanotube can be switched from a conducting state to an insulating state.

A logical switch, basic component of a computer, can be constructed by coupling such CNT transistors together.

(3) *Nanoprobes*

As carbon nanotubes have high flexibility, they can be used in scanning probe instruments. Due to their high conductivity, MWNT tips can be used in AFM and STM machines. Compared with the conventional Si and metal tips, nanotube tips have improved resolutions, and because of their high elasticity, nanotube tips do not crash with sample surfaces. However, because of their high aspect ratio, the vibration of carbon nanotubes is an important issue for nanoprobes.

4.5.2.6 *Nanoelectromechanical systems*

Nanoelectromechanical systems (NEMS) can be defined as nanoscale sensors, actuators and similar devices and systems with critical feature sizes ranging from several hundreds of nanometers to only a few nanometers. The small size of NEMS implies that they have a highly localized spatial response. Moreover, the geometry of a NEMS device can be tailored so that the vibrating element reacts only to external forces in a specific direction. This flexibility is extremely useful for the design of new types of scanning probe microscopes.

The central feature of NEMS is miniaturization, and the characteristics of NEMS obey the scaling laws of the physical world. The effective masses, heat capacities and power consumption are proportional to the critical feature size, either linearly or nonlinearly, while the fundamental frequencies, mass/force sensitivities, and mechanical quality factors are inversely proportional to the critical feature size. Using state-of-the-art surface and bulk nanomachining techniques, NEMS can now be built with masses approaching a few attograms (10^{-18} g) and with cross-sections of about 10 nm. Semiconductor materials such as silicon, silicon carbide and gallium arsenide have been used to build NEMS for a variety of applications.

(1) *Quality factor*

Usually NEMS dissipate very little energy, and this feature is often characterized by the high quality factor (Q) of a resonance. Therefore, NEMS are extremely sensitive to external damping mechanisms, which is crucial for building many types of sensors. In addition, the thermo-mechanical noise, which is analogous to Johnson noise in electrical resistors, is inversely proportional to Q. High Q values are therefore an important attribute for both resonant and deflection sensors, suppressing random mechanical fluctuations and thus making these devices highly sensitive to applied forces.

It should be noted that both intrinsic and extrinsic properties limit the quality factor in real devices. Defects in the bulk materials and interfaces, fabrication-induced surface damages and adsorbates on the surfaces are among the intrinsic features that can dampen the motion of a resonator. The operation of successful NEMS depends largely on defect free structures and devices. Extrinsic effects, such as air resistance, clamping losses at the supports and electrical losses mediated through the transducers, can be reduced by careful engineering. However, some loss mechanisms are fundamental and ultimately limit the maximum attainable quality factors. As we shrink MEMS towards the domain of NEMS, the device physics becomes increasingly dominated by the surfaces. It is expected that extremely small mechanical devices made from single crystals and ultrahigh-purity heterostructures would contain very few defects, so that energy losses in the bulk are suppressed and high quality factors should be possible.

It should be noted that NEMS are intrinsically ultralow-power devices. Their fundamental power scale is defined by the thermal energy divided by the response time, set by Q/w_0. At 300 K, NEMS are only governed by thermal fluctuations when they are operated at the attowatt (10^{-18} W) level. Thus driving a NEMS device at the picowatt (10^{-12} W) scale provides signal-to-noise ratios of up to 10^6. Even if a million such devices were operated simultaneously in a NEMS signal processor, the total power dissipated by the entire system would still only be about a microwatt. This is three or four orders of magnitude less than the power consumed by conventional electronic processors that operate by shuttling packets of electronic charge rather than relying on mechanical elements.

(2) *CNT NEMS*

Because of their unique structures and properties, carbon nanotubes (CNTs) can function as metallic or semiconducting conductors, and have great stiffness in the axial direction. They are excellent thermal conductors, and can readily absorb gases and liquids into their hollow interiors. The combination of these properties makes CNTs natural components for future NEMS. Some of the important properties that are relevant to their use as NEMS device include the nanofluidic behaviors of gases and liquids confined to CNT or bundle interiors, and the mechanical, thermal, and electronic transport properties of CNTs. Both SWNTs and MWNTs are being extensively explored as components in NEMS.

CNT NEMS can be generally grouped into three categories: nanomechanical, resonant and nanofluidic systems. In a CNT nanomechanical system, CNTs typically act as structural components of the system, which deform or move according to external forces. The external forces may be electrostatic, magnetic, thermal, optical, chemical or biological. Important examples of this class of devices include mass/force sensors, torsional springs, nanogears and bearing systems.

In the resonant systems, CNTs are actuated to vibrate by the applied electrical or other signals. Nanoscale resonant systems can be used to detect adsorbed masses. The principle of sensing is based on the resonant frequency shift of a CNT resonator when there is strain developed due to an external loading. In a typical clamped beam, the resonant frequency ν_0 is given by:

$$\nu_0 \approx (E/\rho)^{1/2} \cdot (b/L^2) \tag{4.6}$$

where E is the elastic modulus, ρ is the density, b is the thickness of the beam and L is the beam length. One of the important advantages is that the small effective mass of a nanometer sized beam renders its resonant frequency extremely sensitive to slight changes in its mass. It is found that the frequency can be affected by adsorption of a small number of atoms on the surface.

In the nanofluidic systems, CNTs can function as ion channels and the fluid transport can be controlled by the applied electrical field. The typical examples are the nanoscale drug delivery systems. As the nanofabrication technology is making major advances, the application scene will see many new areas.

(3) *Challenges*

There are three major challenges that must be addressed before the full potential of NEMS can be realized: communicating signals from the nanoscale to the macroscopic world; understanding and controlling mesoscopic mechanics; and developing methods for reproducible and routine nanofabrication.

Apart from these, the development of computational tools to fully design and study NEMS devices is a challenging task. In heterogeneous material systems, the presence of several energy domains such as mechanical, electrical, fluidic, chemical, biological and optical makes the task challenging. Fast and accurate simulation tools are needed for quantum, atomistic, mesoscopic and continuum modeling of the various components of NEMS and their applications in the device.

A large amount of experimental data in the measurement of a variety of macroscopic properties, such as current-voltage relationships and force-displacement curves, need to be validated. Furthermore, the quantum physics problems in the measurement of material properties, surface and interfacial properties, need to be quantified in a reproducible manner. The most critical aspect is the integration of electrical, mechanical, optical and fluidic components on a chip for the NEMS to function. Therefore, a great deal of research efforts are needed for the development of methodologies for integrated nanosystem design, and the development of compact or reduced-order models for the design of nanomechanical, nanofluidic, optical, electronic and other components.

4.5.2.7 *Composite materials*

Because of their stiffness, carbon nanotubes are ideal candidates for structural applications. The Young's modulus of a SWNT could be as high as 1,000 GPa, and that of an MWNT is a little bit lower because the individual cylinders may slide with respect to each other. If carbon nanotubes are used in structural polymer composites, the toughness of the composites will be increased because carbon nanotubes can absorb energy during their highly flexible elastic behaviors. CNT-polymer composites have other advantages, such as low density, increased electrical conduction and better performance during a compressive load.

CNT-polymer composites could be used in many other areas. For example, in the field of biochemistry, CNT-polymer composites could be used as membranes for molecular separation and growth of bone cells.

4.5.2.8 *Templates*

Because of their small channels, carbon nanotubes exhibit strong capillary effects. Due to the strong capillary forces, gases and fluids could be held inside carbon nanotubes. Therefore, nanowires could be created by filling the cavities of carbon nanotubes with appropriate materials. In this way, carbon nanotubes serve as templates.

4.5.2.9 *Neuron growth*

The main requirement of substrates for neuron growth is that they should be light and can be tightly bond with neurons. They can be used for realizing controllable branching of the neuritis and for developing highly directed neuron network which ensure long term cell viability. One important topic in neuroscience research is to look for an amenable way to control neuron growth on nanotube film substrates, such as using electric and magnetic fields.

Carbon nanotubes have been demonstrated as good substrates for neuron growth, and they have two major functions in neuron culturing: guiding neurite outgrowth and regulating neurite branching. Usually, to perform these functions, carbon nanotubes should be specially chemically functionalized. As shown in Figure 4.24, MWNTS are coated with polyethyleneimine (PEI) which promotes neurite outgrowth on the surface [70]. The neurites later form synapses with the neighboring neurons to provide a connection to the neural network.

Figure 4.24 Neurons growing on chemically functionalized MWNTs. Reprinted with permission from Hu, H., Ni, Y., Montana, V., Haddon, R. C. and Parpura, V (2004). Chemically functionalized carbon nanotubes as substrates for neuronal growth, *Nano Letters*, **4** 507–511. © 2004 American Chemical Society.

To explore neurons cultured on nanotubes as biosensors, two issues should be considered. One is that neuron cells should bind specifically and sensitively with odors, drugs and toxins, and the other is that, with the binding of chemicals, the neuron activities should change in response to the chemical substances.

4.5.3 *Carbon nanofoam*

Carbon nanofoams could be used in spintronic devices, which are based on its magnetic properties and semiconductor properties. It could also be used in biomedicine, such as enhancement of magnetic resonance imaging and hyperthermia treatment [5, 6]. The tiny ferromagnetic clusters could be injected into blood vessels to enhance magnetic resonance imaging. Carbon nanofoam could also be implanted in tumors, where it could transfer the energy of external electromagnetic waves into heat, destroying the tumors while leaving the surrounding healthy tissues unharmed.

References:

1. Dai, L. (2006). *Carbon Nanotechnology*, Elsevier Publishing, Netherlands.
2. Rajan, M. S. (2004). *Nano: The Next Revolution*, National Book Trust, New Delhi.

3. Kroto, H. W., Heath, J. R., Obrien, S. C., Curl, R. F. and Smalley, R.E. (1985). C-60 – buckminster fullerene, *Nature*, **318**, 162-163.
4. Iijima, S. (1991). Helical mircotubes of graphic carbon, *Nature*, **354**, 56-58.
5. Rode, A. V., Hyde, S. T., Gamaly, E. G., Elliman, R. G., McKenzie, D. R. and Bulcock, S. (1999). Structural analysis of a carbon foam formed by high pulse-rate laser ablation, *Applied Physics A – Materials Science and Engineering*, **69**, S755-S758.
6. Rode, A. V., Gamaly, E. G., Christy, A. G., Gerald, J. G. F., Hyde, S. T., Elliman, R. G., Luther-Davies, B., Veinger, A. I., Androulakis, J. and Giapintzakis, J. (2004). Unconventional magnetism in all-carbon nanofoam, *Physical Review B*, **70**, 054407.
7. Kohanoff, J., Andreoni, W. and Parrinello, M. (1992). A possible new highly stable fulleride cluster – Li-12C-60, *Chemical Physics Letters*, **198**, 472-477.
8. Smalley, R.E. and Yakobson, B. I. (1998). The future of the fullerenes, *Solid State Communications*, **7** (11), 597-606.
9. Buckminister, F. R. (1984). *The Artifacts of R. Buckminister Fuller: A Comprehensive Collection of His Design and Drawings*, Garland Publishing, New York.
10. Johnson, R. D., Meijer, G. and Bethune, D. S. (1990). C60 has icosahedral symmetry, *Journal of the American Chemical Society*, **112**, 8983-8984.
11. Prinzbach, H., Weller, A., Landenberger, P., Wahl, F., Worth, J., Scott, L. T., Gelmont, M., Olevano, D and von Issendorff, B (2000). Gas-phase production and photoelectron spectroscopy of the smallest fullerene, C-20, *Nature*, **407**, 60-63.
12. Hebard, A. F., Rosseinsky, M. J., Haddon, R. C., Murphy, D. W., Glarum, S. H., Palstra, T. T. M., Ramierz, A. P. and Kortan, A. R. (1991). Superconductivity at 18-K in potassium-doped C-60, *Nature*, **350**, 600-601.
13. Palstra, T. T. M., Zhou, O., Iwasa, Y., Sulewski, P. E., Fleming, R. M. and Zegarski, B. R. (1995). Superconductivity at 40 in cesium doped C-60, *Solid State Communications*, **93**, 327-330.
14. Kozyrev, S. V. and Rotkin, V. V. (1993). Fullerene: structure, crystal-lattice dynamics, electron structure and properties (a review), *Semiconductors*, **27**(9), 777-791.
15. Dresselhaus, M. S., Dresselhaus, G. and Eklund, P. C. (1993). Fullerenes, *Journal of Materials Research*, **8**, 2054-2097.
16. Mitrasinovic, P. and Koruga, D. (1995). Laser-aided production of fullerenes, *Tehnika*, **50**, NM-13.
17. Cortés-Figueroa, J. E. (2003). An experiment for the inorganic chemistry laboratory - The sunlight-induced photosynthesis of (eta(2)-C-60)M(CO)(5) complexes (M = Mo, W), *Journal of Chemical Education*, **80** (7), 799-800.
18. Zimmermann, U., Malinowski, N., Naher, U., Frank, S. and Martin, T. P. (1994). Multilayer metal coverage of fullerene molecule, *Physics Review Letters*, **72**, 3542-3545.
19. Zimmermann, U., Malinowski, N., Burkhardt, A. and Martin, T. P. (1995). Metal-coated fullerenes, *Carbon*, **33**, 995-1006.

20. Martin, T. P., Malinowski, N., Zimmermann, U., Naher, U. and Schaber, H. (1993). Metal-coated fullerene molecules and clusters, *Journal of Chemical Physics*, **99**, 4210-4212.
21. Reynhout, X. E. E., Reijenga, J. C., Notten, P. H. L., Niessen, R. A. H., Daenen, M., de Fouw, R. D., Hamers, B., Janssen, P. G. A., Schouteden, K. and Veld, M. A. J. (2003). *The Wondrous World of Carbon Nanotubes: A Review of Current Carbon Nanotube Technologies*, Eindhoven University of Technology, Netherlands.
22. Niyogi, S., Hamon, M. A., Hu, H., Zhao, B., Bhowmik, P., Sen, R., Itkis, M. E. and Haddon, R. C. (2002). Chemistry of single-walled carbon nanotubes, *Accounts of Chemical Research*, **35**(12), 1105-1113.
23. Damnjanovic, M., Milosevic, I., Vukovic, T. and Sredanovic, R. (1999). Full symmetry, optical activity, and potentials of single-wall and multiwall nanotubes *Physical Review B*, **60** (4), 2728-2739.
24. Gao, B., Kleinhammes, A., Tang, X. P., Bower, C., Fleming, L., Wu, Y. and Zhou, O. (1999). Electrochemical intercalation of single-walled carbon nanotubes with lithium, *Chemical Physics Letters*, **307** (3-4), 153-157.
25. Ding, R. G., Lu, G. Q., Yan, Z. F. and Wilson, M. A. (2001). Recent advances in the preparation and utilization of carbon nanotubes for hydrogen storage, *Journal of Nanoscience and Nanotechnology*, **1**, 7-29.
26. Cumings, J. and Zettl, A. (2000). Low-friction nanoscale linear bearing realized from multiwall carbon nanotubes, *Science*, **289**, 602-604.
27. Ebbesen, T. W. and Ajayan, P. M. (1992). Large-scale synthesis of carbon nanotubes, *Nature*, **358**, 220-222.
28. Bronikowski, M. J., Willis, P. A., Colbert, D. T., Smith, K. A. and Smalley, R. E. (2001). Gas-phase production of carbon single-walled nanotubes from carbon monoxide via the HiPco process: A parametric study, *Journal of Vacuum Science and Technology, A: Vacuum, Surfaces and Films*, **19**, 1800-1805.
29. Wal, R. L. V. and Ticich, T. M. (2001). Flame and furnace synthesis of single-walled and multi-walled carbon nanotubes and nanofibers, *Journal of Physical Chemistry B*, **105** (42), 10249-10256.
30. Wal, R. L. V., Berger, G. M. and Hall, L. J. (2002). Single-walled carbon nanotube synthesis via a multi-stage flame configuration, *Journal of Physical Chemistry B*, **106** (14), 3564-3567.
31. Borowiak-Palen, E., Pichler, T., Liu, X., Knupfer, M., Graff, A., Jost, O., Pompe, W., Kalenczuk, R. J. and Fink, J. (2002). Reduced diameter distribution of single-wall carbon nanotubes by selective oxidation, *Chemical Physics Letters*, **363**, (5-6), 567-572.
32. Huang, S. M. and Dai, L. M. (2002). Plasma etching for purification and controlled opening of aligned carbon nanotubes, *Journal of Physical Chemistry B*, **106** (14), 3543-3545.
33. Gao, B., Bower, C., Lorentzen, J. D., Fleming, L., Kleinhammes, A., Tang, X. P., Mcneil, L. E., Wu, Y. and Zhou, O. (2000). Enhanced saturation lithium

composition in ball-milled single-walled carbon nanotubes, *Chemical Physics Letters*, **327** (1-2), 69-75.

34. Yasuda, A., Kawase, N. and Mizutani, W. (2002). Carbon-nanotube formation mechanism based on in situ TEM observations, *Journal of Physical Chemistry B*, **106** (51), 13294-13298.
35. Jung, S. H., Kim, M. R., Jeong, S. H., Kim, S. U., Lee, O. J., Lee, K. H., Suh, J. H. and Park, C. K. (2003). High-yield synthesis of multi-walled carbon nanotubes by arc discharge in liquid nitrogen, *Applied Physics A – Materials Science and Processing*, **76** (2), 285-286.
36. Anazawa, K., Shimotani, K., Manabe, C., Watanabe, H. and Shimizu, M. (2002). High-purity carbon nanotubes synthesis method by an arc discharging in magnetic field, *Applied Physics Letters*, **81** (4), 739-741.
37. Guo, T., Nikolaev, P., Thess, A., Colbert, D. T. and Smalley, R. E. (1995). Catalytic growth of single-walled nanotubes by laser vaporization, *Chemical Physics Letters*, **243**, 49-54.
38. Yudasaka, M., Yamada, R., Sensui, N., Wilkins, T., Ichihashi, T. and Iijima, S. (1999). Mechanism of the effect of NiCo, Ni and Co catalysts on the yield of single-wall carbon nanotubes formed by pulsed Nd : YAG laser ablation, *Journal of Physical Chemistry B*, **103** (30), 6224-6229.
39. Eklund, P. C., Pradhan, B. K., Kim, U. J., Xiong, Q., Fischer, J. E., Friedman, D., Holloway, B. C., Jordan, K. and Smith, M. W. (2002). Large-scale production of single-walled carbon nanotubes using ultrafast pulses from a free electron laser, *Nano Letters*, **2** (6), 561-566.
40. Maser, W. K., Munoz, E., Benito, A. M., Martinez, M. T., de la Fuente, G. F., Maniette, Y., Anglaret, E. and Sauvajol, J. L. (1998). Production of high-density single-walled nanotube material by a simple laser-ablation method, *Chemical Physics Letters*, **292**, 587-593.
41. Bolshakov, A. P., Uglov, S. A., Saveliev, A. V., Konov, V. I., Gorbunov, A. A., Pompe, W. and Graff, A. (2002). A novel CW laser-powder method of carbon single-wall nanotubes production, *Diamond and Related Materials*, **11**, 927-930.
42. Scott, C. D., Arepalli, S., Nikolaev, P. and Smalley, R. E. (2001). Growth mechanisms for single-wall carbon nanotubes in a laser-ablation process, *Applied Physics A – Materials Science and Processing*, **72**, (5), 573-580.
43. Sinnott, S. B., Andrews, R., Qian, D., Rao, A. M., Mao, Z., Dickey, E. C. and Derbyshire, F. (1999). Model of carbon nanotube growth through chemical vapor deposition, *Chemical Physics Letters*, **315**, 25-30.
44. Lee, C. J., Lyu, S. C., Kim, H. W., Park, C. Y. and Yang, C. W. (2002). Large-scale production of aligned carbon nanotubes by the vapor phase growth method, *Chemical Physics Letters*, **359**, 109-114.
45. Ge, M. H. and Sattler, K. (1994). Bundles of carbon nanotubes generated by vapor-phase growth, *Applied Physics Letters*, **64** (6), 710-711.

46. Lee, J. K., Eun, K. Y., Baik, Y. J., Cheon, H. J., Rhyu, J. W., Shin, T. J. and Park, J. W. (2002). The large area deposition of diamond by the multi-cathode direct current plasma assisted chemical vapor deposition (DC PACVD) method, *Diamond and Related Materials*, **11**, 3-6.
47. Chen, M., Chen, C. M. and Chen, C. F. (2002). Preparation of high yield multi-walled carbon nanotubes by microwave plasma chemical vapor deposition at low temperature, *Journal of Materials Science*, **37** (17), 3561-3567.
48. Huang, Z. P., Wang, D. Z., Wen, J. G., Sennett, M., Gibson, H. and Ren, Z. F. (2002). Effect of nickel, iron and cobalt on growth of aligned carbon nanotubes, *Applied Physics A – Materials Science and Processing*, **74** (3), 387-391.
49. Park, J. B., Choi, G. S., Cho, Y. S., Hong, S. Y., Kim, D., Choi, S. Y., Lee, J. H. and Cho, K. I.. (2002). Characterization of Fe-catalyzed carbon nanotubes grown by thermal chemical vapor deposition, *Journal of Crystal Growth*, **244** (2), 211-217.
50. Su, M., Zheng, B. and Liu, J. (2000). A scalable CVD method for the synthesis of single-walled carbon nanotubes with high catalyst productivity, *Chemical Physics Letters*, **322** (5), 321-326.
51. Zheng, B., Li, Y. and Liu, J. (2002). CVD synthesis and purification of single-walled carbon nanotubes on aerogel-supported catalyst, *Applied Physics A – Materials Science and Processing*, **74** (3), 345-348.
52. Alexandrescu, R., Crunteanu, A., Morjan, R. E., Morjan, I., Rohmund, F., Falk, L. K. L., Ledoux, G. and Huisken, F. (2003). Synthesis of carbon nanotubes by CO_2-laser-assisted chemical vapour deposition, *Physics and Technology*, **44** (1), 43-50.
53. Chiang, I. W., Brinson, B. E., Smalley, R. E., Margrave, J. L. and Hauge, R. H. (2001). Purification and characterization of single-wall carbon nanotubes, *Journal of Physical Chemistry B*, **105** (6), 1157-1161.
54. Farkas, E., Anderson, M. E., Chen, Z. H. and Rinzler, A. G. (2002). Length sorting cut single wall carbon nanotubes by high performance liquid chromatography, *Chemical Physics Letters*, **363** (1-2), 111-116.
55. Hou, P. X., Liu, C., Tong, Y, Liu, M. and Cheng, H. M. (2001). Purification of single-walled carbon nanotubes synthesized by the hydrogen arc-discharge method, *Journal of Materials Research*, **16** (9), 2526-2529.
56. Kajiura, H., Tsutsui, S., Huang, H. J. and Murakami, Y. (2002). High-quality single-walled carbon nanotubes from arc-produced soot, *Chemical Physics Letters*, **364** (5-6), 586-592.
57. Moon, J. M., An, K. H., Lee, Y. H., Park, Y. S., Bae, D. J. and Park, G. S. (2001). High-yield purification process of single-walled carbon nanotubes, *Journal of Physical Chemistry B*, **105** (24), 5677-5681.
58. Chiang, I. W., Brinson, B. E., Huang, A. Y., Willis, P. A., Bronikowski, M. J., Margrave, J. L., Smalley, R. E. and Hauge, R. H. (2001). Purification and characterization of single-wall carbon nanotubes (SWNTs) obtained from the gas-phase decomposition of CO (HiPco process), *Journal of Physical Chemistry B*, **105** (35), 8297-8301.

59. Georgakilas, V., Voulgaris, D., Vazquez, E., Prato, M., Guldi, D. M., Kukovecz, A. and Kuzmany, H. (2002). Purification of HiPCO carbon nanotubes via organic functionalization, *Journal of the American Chemical Society*, **124** (48), 14318-14319.
60. Harutyunyan, A. R., Pradhan, B. K., Chang, J. P., Chen, G. G. and Eklund, P. C. (2002). Purification of single-wall carbon nanotubes by selective microwave heating of catalyst particles, *Journal of Physical Chemistry B*, **106** (34), 8671-8675.
61. Shelimov, K. B., Esenaliev, R. O, Rinzler, A. G., Huffman, C. B. and Smalley, R. E. (1998). Purification of single-wall carbon nanotubes by ultrasonically assisted filtration, *Chemical Physics Letters*, **282**, 429-434.
62. Bandow, S., Rao, A. M., Williams, K. A., Thess, A., Smalley, R. E. and Eklund, P. C. (1997). Purification of single-wall carbon nanotubes by microfiltration, *Journal of Physical Chemistry B*, **101** (44), 8839-8842.
63. Gu, Z., Peng, H., Hauge, R. H., Smalley, R. E. and Margrave, J. L. (2002). Cutting single-wall carbon nanotubes through fluorination, *Nano Letters*, **2** (9), 1009-1013.
64. Goser, K., Glosenkotter, P. and Dienstuhl, J. (2004). *Nanoelectronics and Nanosystems*, Springer, Berlin.
65. Kunzuru, D. and Agarwal, A. (2004). Synthesis and applications of carbon nanotubes, *Directions*, **6** (2), 9.
66. Avouris, P. (2002). Carbon nanotube electronics, *Chemical Physics*, **281** (2-3), 429-445.
67. Ajayan, P. M. and Zhou, O. Z. (2001). Applications of carbon nanotubes, *Carbon Nanotubes*, **80**, 391-425.
68. Zheng, Q. S. and Jiang, Q. (2002). Multiwalled carbon nanotubes as gigahertz oscillators, *Physical Review Letters*, **88**, 045503.
69. Regan, B. C., Aloni, S., Ritchie, R.O., Dahmen, U. and Zettl, A. (2004). Carbon nanotubes as nanoscale mass conveyors, *Nature*, **428**, 924-927.
70. Hu, H., Ni, Y., Montana, V., Haddon, R. C. and Parpura, V (2004). Chemically functionalized carbon nanotubes as substrates for neuronal growth, *Nano Letters*, **4** 507–511.
71. http://en.wikipedia.org/wiki/Carbon_nanotube.
72. Herrera, J., Balzano, L., Pompeo, F., and Resasco, D. (2003). http://www.cmp-cientifica.com/cientifica/frameworks/generic/public_users/TNT02/files/Abstract_Herrera.pdf.
73. www.bristol.ac.uk/depts/chemistry/MOTM/buckyball/c60a.html.
74. www.scielo.br/img/revistas/bjp/v34n4b/a14fig01.gif.

Chapter 5

Nanostructured Materials

Nanostructured materials are extensively investigated and widely used in nanoscience and nanotechnology. In this chapter, typical nanostructured materials will be discussed, including nanopowders, nanoporous materials, nanodusts, nanowires and nanotubes. As an example of three-dimensional nanostructures, zinc oxide nanostructures will also be discussed in details. At the end of this chapter, typical applications of nanostructured materials will be discussed.

5.1 Introduction

The properties and applications of nanomaterials are closely related to the structures of the materials. Due to the varieties in their structures, nanostructured materials can be classified according to different criteria. Usually, nanostructured materials are classified according to their dimensions. The synthesis of nanostructured materials has been a technical challenge in the research and development activities on nanoscience and nanotechnology. Generally speaking, the techniques for synthesizing nanostructured materials fall into two major categories: top-down approach and bottom-up approach.

5.1.1 *Classifications of nanostructured materials*

According to their dimensions, nanostructured materials can be grouped into four broad categories, as tabulated below in Table 5.1 with their characteristics and examples.

Table 5.1 Nanostructured materials and their dimension, characteristics and examples.

Dimension	Characteristics	Examples
0d	These are nanopoint materials. All the dimensions are on nanometric scale preferably less than 3 nm	Quantum dot, nanodust
1d	These are fiberlike nanoparticles. Only two dimensions on nanometric scale	SWNTs, MWNTs
2d	These are plate-like nanoparticles which have only one dimension is on nanomertic scale	Layered silicate (nano clays), nano films
3d	These are equi-axed particles which have all the dimension in nanometric range but more than 4 nm	Nanosilica, nanotitania, nanozirconia

5.1.2 *Top-down and bottom-up synthesis approaches*

Top-down approach and bottom-up approach are two important types of fundamental fabrication approaches for device miniaturization. The ultimate goal is to create the desired properties by designing and building at the nanometer scale. The nanoscale processes in the length scale are schematically shown in Figure 5.1.

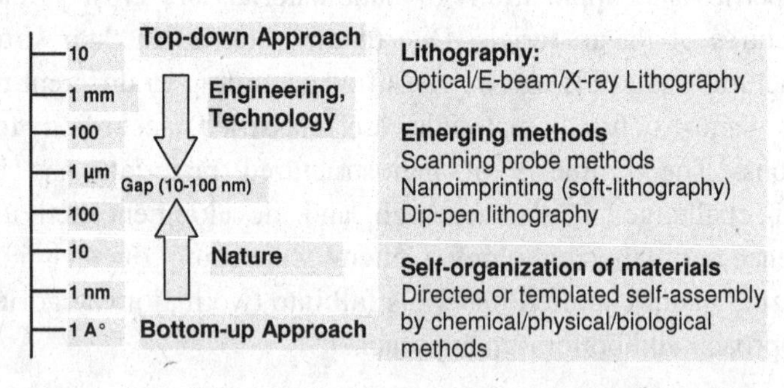

Figure 5.1 Length scale and nanoscale patterning and assembling.

In a top-down technology, nanoscale structures are fabricated by etching and machining techniques. It is a typical technique by which one can chisel away the material to make nanoscale objects. It is almost like a sculptor or a bronze-age scientist working with wood or stone and crafting to the desired shape and size. Attrition or milling is a typical top-down method in making nanoparticles. Some of the examples are photolithography, nanoimprint lithography and nanosphere lithography. However, when a material comes into the nanoscale in a top-down approach, its traditional properties will change, and its surfaces start to dominate the material properties.

Top-down process can also be regarded as a process starting with a bulk material and then breaking it into smaller size using lithography, mechanical, chemical manipulation tools. Optical, electron or ion beams can generate nanoscale patterns over a surface, wherein, these beams can be used as actuators to move atoms or molecules from place to place. Micro tips can be used to emboss or imprint materials and this is a highly controlled process. However, as this process is slow, it is not suitable for large scale production of nanoscale devices.

The biggest problem with top-down approach is the imperfection of surface structure. Top-down techniques such as lithography can cause significant crystallographic damage to the processed patterns, and additional defects may be introduced even during the etching steps. Nanowires, for example, made by lithography is not smooth and may contain a lot of impurities and structural defects on their surfaces. These imperfections may have a significant impact on the physical properties and surface chemistry of nanostructures and nanomaterials. This frequently occurs when the surface over volume ratio in nanostructures and nanomaterials is very large, and these imperfections may result in a reduced conductivity due to inelastic surface scattering. This finally leads to the generation of excessive heat and thus imposes extra challenges to the device design and fabrication. In spite of these anomalies, top-down approaches will continue to play an important role in the synthesis and fabrication of nanostructures and nanomaterials.

In contrast, in a bottom-up technology, organic and inorganic structures are created atom by atom, molecule by molecule or cluster-by-cluster, and this technology is often referred as molecular nanotechnology. This technology synthesizes materials from atomic or molecular species via chemical reactions. In this way, chemical and biological reactors create conditions for special growth and assembly. Any materials containing regular nanosized pores can be used as templates for synthesizing nanoparticles, nanowires or nanotubes. In fact, nanoscale materials and structures are readily formed using bottom-up approaches and materials are self-assembled together to build more complex modules.

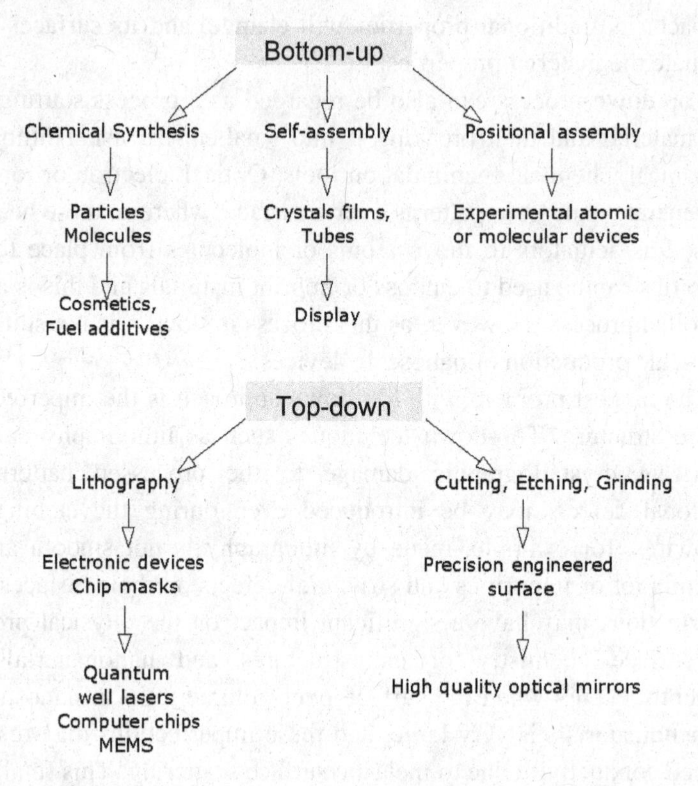

Figure 5.2 Applications of bottom-up approach and top-down approach in engineering.

Figure 5.2 compares the typical applications of bottom-up approach and top-down approach in engineering. Figure 5.3 shows the status of top-down approach and bottom-up approach in engineering [1]. We are at the point where both of the approaches can be used to create new systems, devices and even new materials. As the dimensions that can be controlled by either the top-down approach or the bottom-up approach are of a similar order, a new approach, hybrid approach, is often used. Working around 100 nm dimensions, the lithography may be considered as a hybrid approach, since the thin films are usually fabricated in a bottom-up approach, whereas etching is a typical top-down approach.

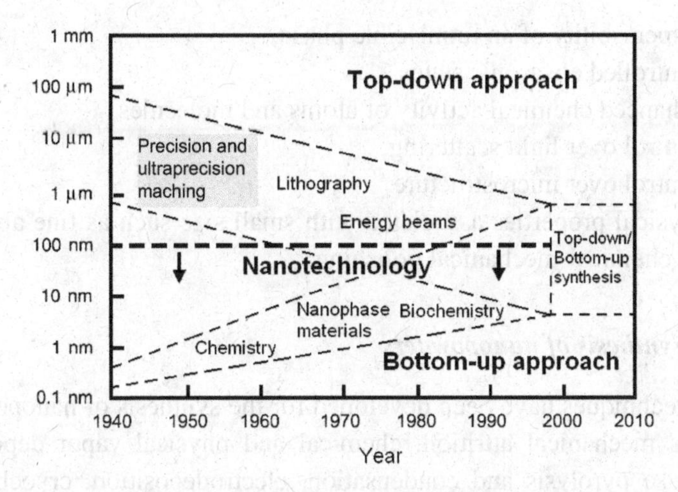

Figure 5.3 Status of top-down and bottom-up approaches in nanotechnology [1].

5.2 Nanopowders

Nanopowders are particulate materials with dominant structural features on size scales from 1 to 100 nm. Some kinds of nanopowders, such as fumed silica and carbon black, can be industrially produced in tons, while some kinds of nanopowders, such as stabilized metallic quantum dots and atomic clusters, are extremely precious, and are available in only sparing quantities. In the following, we discuss the typical properties of nanopowders and the techniques often used in the synthesis of nanopowders.

5.2.1 *Properties of nanopowders*

Compared with those in bulks, the atoms and molecules at the surfaces of nanopowders are at higher energetic states. The higher energy associated with the surface make nanopowders exhibit special properties, mainly including:

- Uniformity and fine grain size,
- Extremely high surface area per mass,
- Transport properties of small domains and pores,
- Absorptive capabilities of fine particles,

- Dispersability of an immiscible phase,
- Controlled electronic states,
- Enhanced chemical activity of atoms and molecules,
- Control over light scattering,
- Control over microstructure,
- Physical properties associated with small size such as fine abrasives for chemical-mechanical polishing.

5.2.2 *Synthesis of nanopowders*

Many techniques have been developed for the synthesis of nanopowders, such as mechanical attrition, chemical and physical vapor deposition, gas phase pyrolysis and condensation, electrodeposition, cryochemical synthesis and sol-gel methods. In the following, we concentrate on the synthesis of two typical types of nanopowders: metal nanopowders and metal oxide nanopowders.

5.2.2.1 *Metal nanopowders*

Nanostructured metal powders may be produced either via the reduction or co-reduction of metal salts using alkaline-triorganohydroborates or using the "polyol" or the "alcohol-reduction" pathways.

(1) *Triorganohydroborate reduction*

Pt nanopowders can be obtained from Pt-salts by triorganohydroborate reduction. The Pt nanopowders obtained in this way have an average size of 3-4 nm with purities higher than 95%. However, they have a broad size distribution, are often contaminated with small residues of alkaline halides.

(2) *Polyol method*

This method is based on the decomposition of the ethylene glycol. The Pt nanopowders obtained by this method have sizes in the range of 5-13 nm, and their purities are higher than 99%.

(3) *Alcohol reduction method*

Alcohols, such as methanol, ethanol or propanol, can be used as reducing agents: they get oxidized to aldehydes or ketones. By refluxing metal salts or complexes, such as H_2PtCl_6, $HAuCl_4$, $PdCl_2$, $RhCl_3$, in an alcohol/water (1/1 by volume) solution, nanocrystalline metal powders can be obtained in the absence of stabilizers.

5.2.2.2 *Metal oxide nanopowders*

The typical methods for synthesizing metal oxide nanopowders include sol-gel method and hydrothermal/solvothermal method.

(1) *Sol-gel method*

This technique is different from the traditional process fusion. In the preparation of the "sol", the starting materials are usually inorganic metal salts or metal organic compounds such as metal alkoxides. The precursor is subjected to a series of hydrolysis and polymerization reactions to form a colloidal suspension (sol):

M-O-R + H_2O = M-OH + R-OH (Hydrolysis) (5.1)
M-OH + HO-M = M-O-M + H_2O (Water condensation) (5.2)
M-O-R + HO-M = M-O-M + R-OH (Alcohol condensation) (5.3)

with M = Si, Zr, Ti.

The precursors used in sol-gel methods are not very reactive despite other precursors for oxidic materials such as titania, tin oxide, etc. This makes it easy to control the silica sol-gel process. Alkoxysilane compounds are used in most cases. Usually, the solid particles suspended in sols have a diameter of a few hundred nanometers. An overview of the chemicals and conditions for the preparation of the various sols is tabulated below in Table 5.2.

The sol condenses in a new phase, gel, in which solid macro-molecules are immersed in a liquid phase. By casting the "sol" into a mold, a wet "gel" will form. A gel can be modified with a number of dopants to achieve unique properties. Ultra-fine and uniform ceramic powders can be formed by precipitation, spray pyrolysis or emulsion techniques.

Table 5.2 Preparation of various sols.

Sol	Precursors	Solvents/other chemicals	pH
$BaTiO_3$	titanium (IV) isopropoxide, barium acetate	glacial acetic acid, ethylene glycol	~4
ITO	indium chloride	ethylene glycol, ethanol, citric acid, water tin (IV) chloride	~1
Nb_2O_5	niobium chloride	ethylene glycol, ethanol, citric acid, water	~1
PZT	lead (II) acetate, titanium isopropoxide, zirconium n-propoxide	glacial acetic acid, water, lactic acid, glycerol, ethylene glycol	~4
SiO_2	tetraethyl orthosilicate	ethanol, water, hydrochloric acid	~2
$SrNb_2O_6$	strontium nitrate, niobium chloride	ethylene glycol, ethanol, citric acid, water	~1
TiO_2	titanium (IV) isopropoxide	glacial acetic acid, water	~2
V_2O_5	vanadium pentoxide	hydrogen peroxide, water, hydrochloric acid	~2.7

(2) *Hydrothermal/solvothermal process*

In this process, solvents can be brought to temperatures well above their boiling points, in a sealed vessel, such as autoclave, by the increase in autogenous pressures resulting from heating. In a solvothermal process, a chemical reaction under high temperature and high pressure takes place whereas in a hydrothermal process, water is used as a solvent. At high temperature and pressure, solvents exhibit high viscosities and easily dissolve chemical compounds that would otherwise exhibit very low solubilities under ambient conditions. Hydrothermal/solvothermal processing allows many inorganic materials to be prepared at temperatures substantially below those required by traditional solid-state reactions.

5.3 Nanoporous Materials

The field of nanoporous materials is very interdisciplinary. It ranges from a variety of network compositions (inorganic, organic and metalloorganic), different pore shapes (disordered, spherical, cylindrical, lamellar), different pore sizes (from 0.5 nm to several tenths of nanometers) and different pore surface properties [2]. Classically, a porous material is a material that has through voids. The voids show a translational repetition in a three-dimensional space, while no regularity is necessary. A typical and relatively simple porous system described in colloid science is a solid foam. In correlation with this, the most dominant characteristic of a porous material is its gas-solid interfaces.

The major parameters for nanoporous materials are size of pores, type of network material and state of order. Nanoprous materials may be inorganic, such as silica, nitride and $AlPO_4$, or organic, such as charcoal.

It should be noted that most of the porous materials are not stable by thermodynamic means. As soon as kinetic energy boundaries are overcome, materials start to break down. Porous silica, for example, is just metastable. As soon as the temperature is raised and the melting point is reached, primary particles in the network begin to fuse. Finally, at very high temperatures, the thermodynamic stable phase of SiO_2 quartz emerges. Control over interface energy and metastabilization of nanodimensional holes is of special importance for the production of nanoporous materials.

5.3.1 *Properties of nanoporous materials*

Nanoporous materials have pore sizes in the range between 2 nm and 50 nm. If a high porosity (larger than 50%) can be achieved, the high porosity at the nanoscale is always equivalent to a high surface area. For instance, a material that has 80% spherical 5-nm voids can theoretically contribute a 480 m^2/g surface area.

In an ordered nanoporous material, the pores are monodisperse in size, and have a specific shape and a mutual 3D correlation between the pores [3]. Both of the ordered nanoporous materials and disordered nanoporous materials are very important for practical applications, such as sensing, catalysis, dielectric coatings and molecular sieving.

5.3.2 *Synthesis of nanoporous materials*

One can find a whole variety of nanoporous materials in nature executing many different functions. Even in the human body, life without bones would not be conceivable. It is seen that complex mechanisms are involved in the formation of these hierarchical materials. Similar to the structure motives on different length-scales cells, vesicles, supramolecular structures and biomolecules are involved in the structuring process of inorganic matter occurring in nature. This process is commonly known as biomineralization, and it has been founded that ordered porous materials, and therefore artificial materials, are constructed according to very similar principles. A completely different area where nanoporous materials are highly important is in the lungs, where a foam with a high surface area permits sufficient transfer of oxygen to the blood.

Even the most recent developments in nanoporous materials, such as their application as photonic materials, are already present in nature. The color of butterfly wings, for instance, originates from photonic effects. We can say that nature applies the concept of nanoporous materials as a powerful tool for constructing all kinds of materials with advanced properties.

The most successful way to produce nanoporous materials is the templating method. An organic, or sometimes inorganic, compound acts as a place holder that later becomes a void space in nanoporous materials. The templating concept obviously allows control on pore size and shape. Initially, a suitable template structure, which must be compatible with solvents involved in the process and the final network materials, has to be provided. Macrophase separation must be avoided at all times [4].

Template method has been used to synthesize various materials, such as metal, polymer, ceramics. The as-prepared materials are usually in fiber or tubular form with micron or nano sizes. Anodic aluminum oxide (AAO) and track etch membranes are two types of templates which are commonly used. Both of them have a packed array of columnar cells with uniformly sized holes whose diameter is in the range of 4 to 400 nm.

The templating concept can be extended toward pure organic or polymeric materials. In this way, organic monomers are polymerized around the prefabricated template structures. The chemical and physical parameters of the polymers obtained in this way are quite different from those of the monomer species. One important characteristic of this approach is that restructuring processes occur during the polymerization procedure.

The interactions between the template phase and the network builder are of crucial importance. Furthermore, entropic forces such as restriction of polymer segment mobility due to spatial restrictions can become important. These factors decide whether a synthesis is transcriptive or reconstructive. One problem that also remains is that the porous materials are just metastable. This problem becomes much more severe for polymeric materials than for most inorganic networks. Because of swelling of polymers and their high flexibility, polymer systems might be unable to retain the porous structure in the long run.

In the track-etch method, a nonporous sheet of the desired material, such as polycarbonate, polyester and PET, is bombarded with nuclear fission fragments to create damage tracks in the sheet, and these tracks are then chemically etched into pores. The size, shape and density of the pores can be controlled to achieve desired transport and retention characteristics.

Various kinds of nanoporous materials have been produced. In the following, we discuss several typical kinds of nanoporous materials.

5.3.2.1 *Silica*

Porous materials, such as activated carbon, silica gel and zeolite are widely used as adsorbents. Porous materials with micro pores can be readily produced by a variety of techniques, but the generation of "mesopore" material is more difficult [4, 5]. Porous silica with nm-sized pores are generally prepared from fine silica powder, by

- Acid deposition from Na_2SiO_3 solution
- Sol-gel method, from organo-silicone compounds

- Vapor deposition from a silica fume
- Hydrothermal process

Porous silica with controlled pore size can be produced by hydrothermal process and using calcium silicate as the starting material. Silica aerogels and xerogels are porous materials, which possess many exceptional properties. Highly volatile liquids like ethanol or isopropanol (IPA) have been used for the solvent in the traditional two-step acid-base sol-gel process, which may deteriorate the reproducibility of the process. It is also demonstrated that aerogel films act as a source for particles during the supercritical drying process.

Ordered nanoporous silica materials can be obtained, which have a variety of pore sizes (2-80 nm), a variety of pore shapes (from spheres to lamellae), and a variety of surface properties. This makes ordered mesoporous silica materials a very active field of research.

The density of porous silica thin film may be measured by grazing incidence X-ray reflectivity (GIXR). The pore size distribution of cure samples can be measured by grazing incidence small angle X-ray scattering (GISAXS).

5.3.2.2 *Transition metal oxides*

Because of their extensive applications in many areas, such as catalysts and redoxes, much work has been devoted to the synthesis of stable ordered mesoporous transition metal oxides. Many compositions with Ti, Zr, V, Nb, Ta, Mo, W, Mn and Y as the central elements have been fabricated. Because of its potential use in solar cells and electrochromic devices, special attentions have been paid on ordered mesoporous titania, a quite representative example for transition metal-based materials. The reactivity of titania precursors goes in the order of $TiCl_4 \rightarrow Ti(iOPr)_4 \rightarrow Ti(OBu)_4 \rightarrow Ti(OEt)_4$.

5.3.2.3 *Metal sulfides*

Mesostructured metal sulfide materials can be developed through a true liquid crystal approach. In the fabrication procedure, a lyotropic block copolymer phase is swollen by an inorganic salt solution, such as Cd_2C salts. After treatment with H_2S vapor, a nanocrystalline, ordered mesoporous CdS material can be obtained.

5.3.2.4 *Aluminum phosphates*

Because of the similarity between silica molecular sieves (zeolites) and aluminum phosphates (ALPOs), many attempts have been made to synthesize MCM-41 analogous ALPO materials. In most cases, nonstable materials are obtained, or the pore systems are quite different from MCM-41. Kuroda *et al.* reported a nicely ordered hexagonal ALPO material in 1999 [6].

5.2.3.5 *Silicon nitrides*

Just like ALPOs, it is possible to produce an ordered mesostructured silicon nitrides because the NH unit in imides or amines is isoelectronic to oxygen, which would allow a sol-gel chemistry similar to that of silica. Attempts have been made for synthesizing these compounds with silicon amines, ammonia and nonaqueous solvents and surfactants.

5.3.2.6 *Anodic aluminum oxides*

An anodic aluminum oxide (AAO) membrane can be formed on an aluminum strip by continuously passing the strip through a phosphoric acid electrolyte. The film may be made by anodizing one or more aluminum metal anodes of suitable area in an electrolyte based on phosphoric acid at a concentration of 5-150 g/l maintained at a temperature in the range 5-50°C using an anodizing voltage of 40-200 V. In the resulting film, the pores are enlarged so that the effected surface area is increased. The film forms an excellent substrate for lacquer, paint or adhesive. After separation from the metal substrate, the films have interesting properties for use in filters.

Ordered mesoporous alumina would also be a very interesting material because of its high relevance in catalysis. The classic way of preparing mesoporous alumina is electrolytic etching (anodized alumina), and the results obtained are impressive for the production of thin films. This technique is restricted to surfaces [7]. Therefore, preparing ordered mesoporous alumina via templating routes is very promising. It is very difficult to achieve highly ordered materials and to achieve control over pore size with alumina, although some sucessful examples exist.

AAO membranes are macroscopically highly ordered systems made of oriented nanoscale parallel cylindrical channels. The membrane topology is defined by the pore radius, the inter-pores distance and the pore length. The morphology of membrane can be tailored by the main anodization parameters: voltage, temperature and anodization duration.

The crystal structure of the AAO membrane can be determined by using X-ray diffraction (XRD) and transmission electron microscopy (TEM) [8, 9]. A scanning electron microscopy (SEM) can be used to estimate the pore area on the surfaces of membrane. The surface roughness on the membranes can be evaluated by atomic force microscopy (AFM), and the root-mean-square roughness values can also be obtained.

5.3.2.7 *Metals*

Hexagonally ordered metal sponges may be prepared by using a templating technique. This technique is very similar to the ones described above, although the network is now formed by the reduction of noble-metal precursors.

5.4 Nanodusts

Nanodusts are a type of nanopoint materials with all the three dimensions preferably less than 3 nm. Nanodusts can be artificially synthesized, and can also be found in nature.

5.4.1 *Nanodusts in nature*

Nanodusts exist in nature. Figure 5.4 shows the nanodust particles attached to the legs of honeybee. Besides, it has been found that nanoscale structures in tornadoes can produce a rapid charge of electricity, sending thunderclouds into a spin [10, 11].

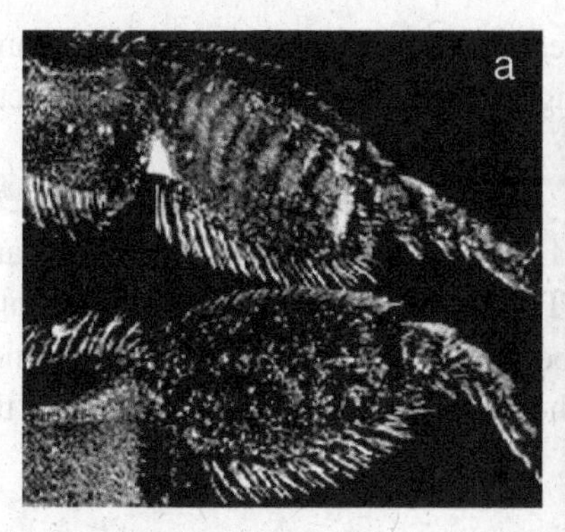

Figure 5.4 (a) Microscopic view of the legs of a honeybee; (b) attached nanodust particles and pollen on the legs of honeybee. (Image courtesy of Jeff Brinker, Sandia National Laboratories)

5.4.2 *Synthesis of nanodusts*

Many methods have been developed for producing nanodusts [12-14]. In the choice of production methods, following factors should be taken into consideration: desired particle size and particle size distribution, required structure and purity, availability of raw materials, desired production rate and cost of production.

Crushing ceramics and refractory materials in orbital mills is a cost effective method to produce nanomaterials, including nanodusts. The self-abrasive method is another cost effective method and it provides uniform size distribution. In this method, a gas jet captures the particles and, with a speed near to that of sound, carries them upwards, to the separation zone. A thin fraction, or the smallest of them, is separated by the centrifugal separator. At the same time the heavy and large particles fall back again, into the grinding zone. Owing to the particles hitting each other, they break down into pieces and, and they are also polished due to friction. These methods are very convenient for making various refractory elements, such as turbine blades, and they are also not prohibitively expensive.

Silicon nanodusts can be formed by chemical or electrochemical etching of boron-doped silicon that is anodized in solutions of H_2O_2 and HF. Under these circumstances small islands are formed at meniscus-like area. These islands or large particles are then crushed in an ultrasound bath and nanodusts of silicon having dimensions near 1 nm in diameter are formed.

Silicon nanodusts can also be synthesized by laser-induced decomposition of SiH_4, because SiH_4 is very volatile pyrophoric gas and it combusts spontaneously [15]. This technique requires the photo-decomposition of SiH_4 and the liberation of gaseous Si atoms which would collide and recombine with other gaseous Si atoms until the required nanodusts are made.

5.5 Nanowires

A nanowire is a structure that has two lateral sizes constrained to tens of nanometers or less and one unconstrained longitudinal size, and it is often referred to as a type of one-dimensional material [16, 17]. In most cases, a nanowire has a circular or elliptic cross section. As its diameter is at the nanoscale, a nanowire is often called a quantum wire, due to its quantum mechanical effects. Nanowires have many interesting properties that are not seen in three-dimensional bulk materials. This is because the electrons in nanowires are confined laterally and thus occupy the energy levels different from the traditional energy levels in bulk materials.

5.5.1 *Classifications of nanowires*

Nanowires can be classified in many ways. According to their materials, nanowires can be generally classified into five categories, including metallic nanowires (e.g., Ni, Pt, Au), semiconducting nanowires (e.g., Si, InP, GaN), dielectric nanowires (e.g., SiO_2, TiO_2), magnetic nanowires (e.g., Ni, Co, Fe_2O_3) and molecular nanowires composed of repeating molecular units either organic (e.g., DNA) or inorganic (e.g., $Mo_6S_{9-x}I_x$).

As shown in Figure 5.5, according to their structures, nanowires can be classified into three major categories: single-component nanowires, multi-component nanowires and multi-layer nanowires.

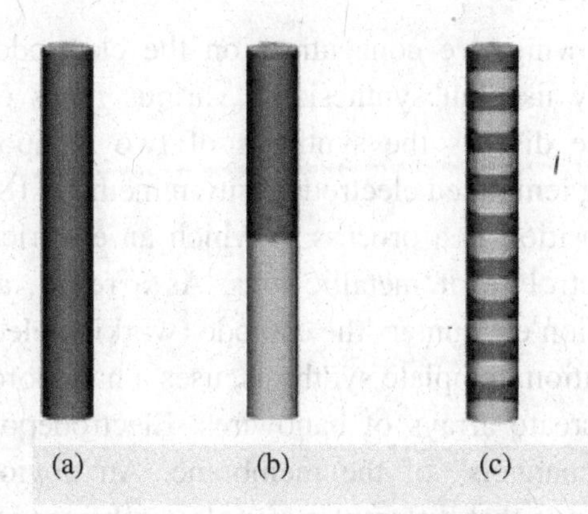

Figure 5.5 Schematic illustration of three types of nanowires [17]. (a) Single-component nanowire; (b) two-component nanowire with segment aspect ratio >1; (c) two-component multi-layer nanowire with segment aspect ratio <1.

5.5.2 *Synthesis of nanowires*

As nanowires could not be synthesized spontaneously in nature, many efforts have been made in synthesizing nanowires for various purposes, and as a result, many methods have been developed for the synthesis of nanowires with different specifications. Table 5.3 lists the methods often used for synthesizing nanowires [16].

Table 5.3 Methods often used for synthesis of nanowires [16].

Materials	Synthetic methods	Dimensions: diameter d, length l	Aspect ratio
Ni, Au, Pt, Ag, Co, Cu, ZnO, conductive polymers, more	Templated electrodeposition	d = 10-350 nm, l = up to 50 mm	up to 250
TiO_2, ZnO	Sol–gel	d = 10-350 nm, l = up to 50 mm	250
Silicon	Nanocluster-mediated vapor–liquid–solid growth	d = 10 nm, l = >1 mm	>100
MoO_2, Fe_2O_3, Cu_2O, Pd[108]	Electrodeposition on graphite surface	d = 13 nm and up, l = microns	>100
Cu, Au, and Ag	Electrodeposition on graphite surface	d = 50 nm, l = microns	>20

In the following, we concentrate on the electrodeposition method which is widely used in synthesizing various types of nanowires. As an example, we discuss the synthesis of two component nickel-gold nanowires using templated electrodeposition method [18].

Electrodeposition is a process in which an electrical current passes through an electrolyte of metallic ions. As a result, a reduction takes place when the ion encounters the cathode (working electrode). One type of electrodeposition, template synthesis, uses a nanoporous membrane as a structure to create arrays of nanowires. Electrodeposition then takes place in the "channels" of the membrane. An obvious advantage of electrodeposition is that different materials can be sequentially deposited in the templates, and so nanowires comprised of segments of different materials can be synthesized using this method.

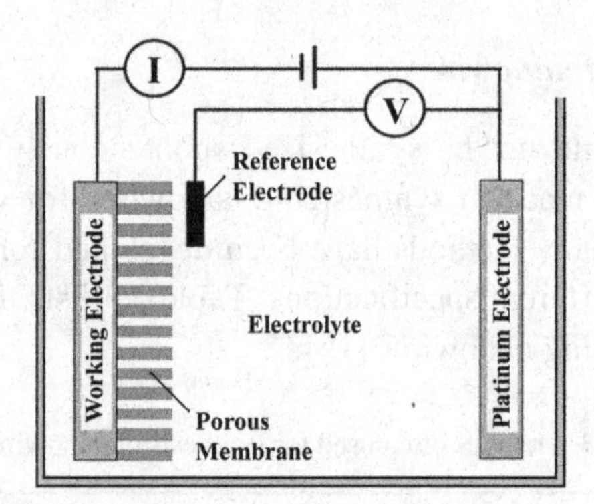

Figure 5.6 Experimental setup for synthesis of nanowires. (With kind permission from Springer Science + Business Media: Bera, D., Kuiry, S. C. and Seal, S. (2004). Synthesis of nanostructured materials using template-assisted electrodeposition, *JOM*, **56**(1), 49–53, Figure 6.)

Electrodeposition of nanowires can be done in a three-electrode arrangement as shown in Figure 5.6 [19]. A thin layer of copper film sputter-deposited on the alumina membrane serves as the working electrode with a platinum counter electrode and an Ag/AgCl reference electrode. Prior to the deposition of nanowires, a small amount of copper is grown in the pores to reinforce the sputtered copper layer. Copper is deposited from a $CuSO_4 \cdot 5H_2O$ aqueous solution. Nickel is deposited from a $NiCl_2 \cdot H_2O$, $Ni(H_2NSO_3)_2 \cdot 4H_2O$ and H_3BO_3 aqueous solution.

Gold is deposited from a gold electrodeposition solution. Two component nickel-gold nanowires can be prepared by first depositing gold, changing the solution, and then depositing nickel.

After electrodeposition of the nanowires, the copper layer can be removed with a solution of $CuCl_2$ in HCl, and the alumina membrane can then be dissolved in KOH solution. After that single nickel-gold nanowires can be collected from solutions with a magnet or by centrifugation. The supernatant is decanted and replaced with neat solvent. The nanowires are then sonicated in an ultrasonicator. This collection/solvent replacement/sonication process is usually performed several times. Finally, to realize special functions, functional groups are selectively functionalized on the surfaces of the nanowires. Figure 5.7 shows a piece of nickel-gold nanowire fabricated in this method.

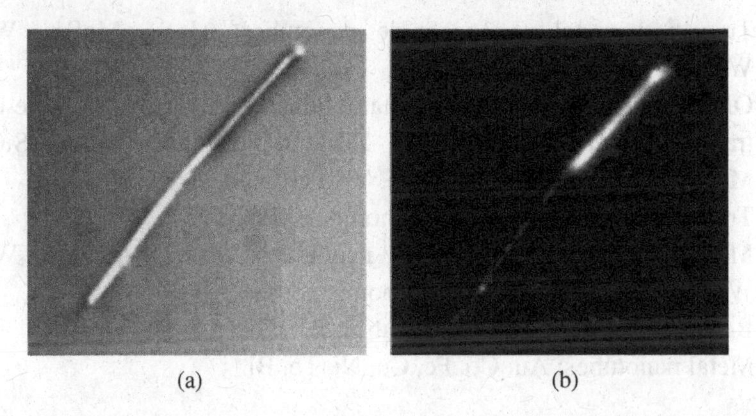

Figure 5.7 A piece of 22-μm nickel-gold nanowire functionalized with HemIX and nonylmercaptan. (a) Reflection image of a nanowire and (b) fluorescent image of the nickel segment. Reprinted with permission from Bauer, L. A., Reich, D. H. and Meyer, G. J. (2003). Selective functionalization of two-component magnetic nanowires, *Langmuir*, **19**, 7043-7048. © 2003 American Chemical Society.

5.6 Nanotubes

Similar to nanowires, nanotubes are a type of high aspect ratio nanomaterials; however, the most obvious structural difference between nanowires and nanotubes is that usually a nanowire is solid, while a nanotube is void inside, with distinctive inner and outer surfaces.

5.6.1 *Classification of nanotubes*

Nanotubes can be classified into two major categories: carbon nanotubes and inorganic nanotubes. Although there are some geometrical similarities to carbon nanotubes, inorganic nanotubes are distinguished by important peculiarities, from their growth mechanisms to the physical and chemical properties, which are attractive for potential applications [20, 21]. The properties, synthesis methods and applications of carbon nanotubes are discussed in Chapter 4. Here we concentrate on inorganic nanotubes.

Since the first report on inorganic WS_2 nanotubes in 1992 [22], many efforts have been made on synthesizing inorganic nanotubes. The inorganic nanotubes that have been reported can be classified into following six major families [20]:

- Transition metal chalcogenide nanotubes: MoS_2, $MoSe_2$, WS_2, WSe_2, NbS_2.
- Oxide nanotubes, including metal oxides (α-Fe_2O_3, γ-Fe_2O_3, Fe_3O_4), transition metal oxides (TiO_2, ZnO, VO_x), silicon oxide (SiO_2), MoO_3, RuO_2, rare earth (Er, Tm, Yb, Lu) oxides.
- Transition metal halogenous nanotubes: $NiCl_2$.
- Mixed-phase and metal-doped nanotubes: $PbNb_nS_{2n+1}$, $Mo_{1-x}WS_2$, $W_xMo_yC_zS_z$, Nb-WS_2, WS_2-carbon nanotubes.
- Boron- and silicon-based nanotubes: BN, BCN, Si.
- Metal nanotubes: Au, Co, Fe, Cu, Ni, Te, Bi.

5.6.2 *Synthesis of inorganic nanotubes*

The methods for growing inorganic nanotubes generally fall into six types: sulfurization, decomposition of precursor crystals, template growth, precursor-assisted pyrolysis, misfit rolling, direct synthesis from the vapor phase. Table 5.4 lists typical examples of the use of each type of method. By affecting the growth rate and the available energy for the structural relaxation, each growth technique has a particular effect on the nanotube morphology. It should be noted that in some cases, nanotubes are grown by combining several growth techniques.

Table 5.4 Growth methods for various inorganic nanotubes [20].

Compound	Method	Synthetic route
WS_2, MoS_2	Sulfurization	Heating MoO_x or WO_x in H_2S
ZnS		Heating ZnO in H_2S
NbS_2, TaS_2, HfS_2, ZrS_2, VOx	Decomposition of precursor crystals	Oxidation of tri and tetra transition metal chalcogenides at elevated temperatures
TiO_2, $AlsO_3$ Au, CO, Fe, Si GaN	Template growth	Sol–gel, electrodeposition
Vo_x	Layered precursors	Solvothermal synthesis
Bi	Lamellar precursors	Hydrothermal pyrolysis
INGaAs/GaAs	Misfit rolling	Rolling up of strained heterostructures
MoS_2, WS_2, Au-MoS_2, $PbNb_nS2_{n+1}$	Direct synthesis from vapor phase	Chemical transport reaction

As an example, we discuss the synthesis of iron oxide nanotubes using thermal decomposition method [23, 24]. To form nanotubes, alumina templates are used as substrates to provide an array of pores. To obtain iron oxide nanotubes, the substrates are wetted with alcohol and loaded with a high concentration (>60 wt%) $Fe(NO_3)_3 \cdot 9H_2O$ in alcohol solution at room temperature. This solution is then forced into the template pores at room temperature using a pressure cell. The loaded substrates are cleaned, mounted on a sample holder, and placed in the oven with pores horizontal. By heating the loaded substrates in air at a rate of 10°C/min to 250°C, and stay at this temperature for 4 hours, hematite magnetic nanotubes can be obtained due to the decomposition of the iron nitrate. To obtain magnetic nanotubes, the substrates are further reduced at the same temperature for 4 hours in flowing hydrogen and subsequently cooled to room temperature. After the nanotubes are formed, samples are etched in a NaOH aqueous solution, and the precipitates are dispersed in acetone. Figure 5.8 shows the dispersed iron-oxide nanotubes fabricated using this method.

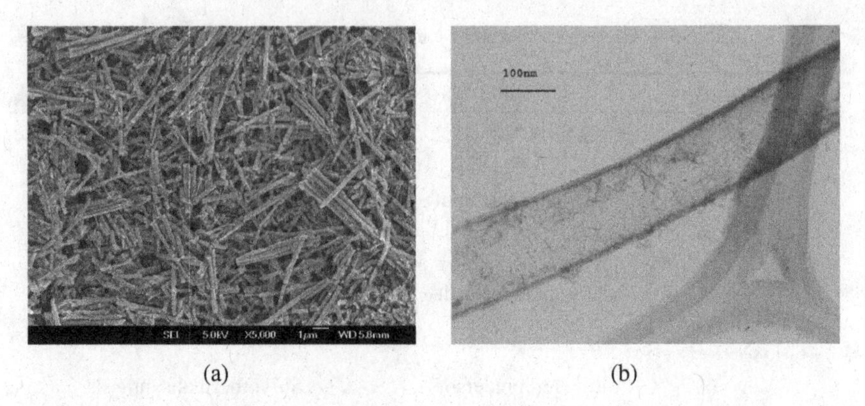

Figure 5.8 Iron-oxide nanotubes fabricated using thermal decomposition method. (a) SEM picture. (b) TEM picture. [24]

5.7 Zinc Oxide Nanostructures

Because of their remarkable physical properties, ZnO nanomaterials have been proposed to be a more promising UV emitting phosphor than GaN because of their large exciton binding energy of 60 meV, which is about three times as large as that of GaN. This leads to a reduced UV lasing threshold and yields higher UV emitting efficiency at room temperature. The interesting properties of nanowires are mainly due to the discrete energy level spacing at the edges of conduction and valence bands. As the size of a nanowire becomes smaller, the highest occupied molecular orbit-lowest unoccupied molecular orbit (HUMO-LUMO) gap becomes larger, resulting in the blue shift of the UV-visible absorption spectra.

Many efforts have been made on the synthesis, characterization and applications of ZnO nanomaterials. An assortment of ZnO nanostructures, such as nanorods, nanotubes, nanorings and nano-tetrapods, have been successfully grown by a variety of methods, such as chemical vapor deposition, thermal evaporation, hydrothermal technique, sol-gel and electrodeposition [25-34]. However, the yields of ZnO nanostructures are usually quite low.

Here, we discuss a novel synthesis method that has much higher productivity [34]. In this method, ZnO nanostructures are synthesized from zinc nitrate hexahydrate ($Zn(NO_3)_2$) and hexamethyltetramine

(HMTA), and the synthesis procedure consists of two parts. In the first part, 400 ml of 0.1M $Zn(NO_3)_2$ and 400 ml of 0.2M HMTA solutions are prepared and the solutions are stirred for 30 minutes. Then the $Zn(NO_3)_2$ solution is placed on a hot plate and heated to 90°C under constant stirring. After that, the HMTA solution is introduced drop by drop into the $Zn(NO_3)_2$ solution, and the resulting solution is allowed to stand on the hot plate for 60 minutes under constant stirring. Then the solution is allowed to cool down for 10 minutes. A white precipitate is observed in the bottom of the beaker and it is filtered out using filter paper. The precipitate is kept in a gravity oven at 100°C for 24 hours and white powders are obtained as shown in Figure 5.9.

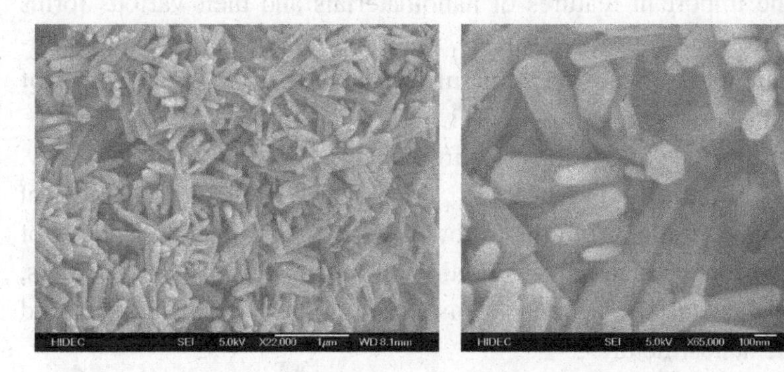

Figure 5.9 SEM pictures of the precipitate obtained in the first part.

Figure 5.10 SEM picture of the precipitate obtained in the second part.

In the second part, all the steps in the first part are repeated, and the only additional step is the addition of 0.1 gms of the powders obtained in step 1 into the $Zn(NO_3)_2$ solution before the addition of the HMTA solution. The resultant precipitate is shown in Figure 5.10. It is found that the secondary ZnO nanostructures seem to grow around the primary ZnO nanocrystals synthesized in the first part. In both parts, the total yield is close to 5 gms. This kind of yield is essential for applications which require ZnO nanocrystals in bulk quantities.

5.8 Applications

Due to the important features of nanomaterials and their various forms such as powders, arrays, particles, crystals, single-walled and multi-walled nanotubes, nanostructured materials have a wide spectrum of applications, as shown in Figure 5.11.

Generally speaking, nanostructured materials can be used in energy, electronics, optical engineering, chemical industry, biomedical engineering and defense. In the following, we discuss the applications of the typical nanostructured materials discussed above: nanopowders, nanoporous materials, nanodusts, nanowires, inorganic nanotubes and zinc oxide nanostructures.

5.8.1 *Nanopowders*

Nanosized powders have been used for various applications, such as cosmetics, anti-microbial products, sunscreen and electro catalysts for fuel-cells. Some of the specific applications are listed below:

- Zinc oxide: cosmetics, electronics and pigments.
- Cupric oxide (copper (II) oxide): electronics, catalysis and antimicrobial applications
- Magnetite (iron II(III) oxide): magnetic applications, coatings, cosmetics and pigments.
- Stannic oxide (tin (IV) oxide): batteries, catalysis, electronics and sensors.
- Anatase (titanium (II) oxide): cosmetics, pigments, coatings and photocatalytic applications.

- Nano platinum metal supported on activated carbon: electrocatalyst for fuel cells and metal-air batteries.
- Nano silver metal supported on activated carbon: electrocatalyst for fuel cells and metal-air batteries, water treatment, medical and anti-microbial applications.

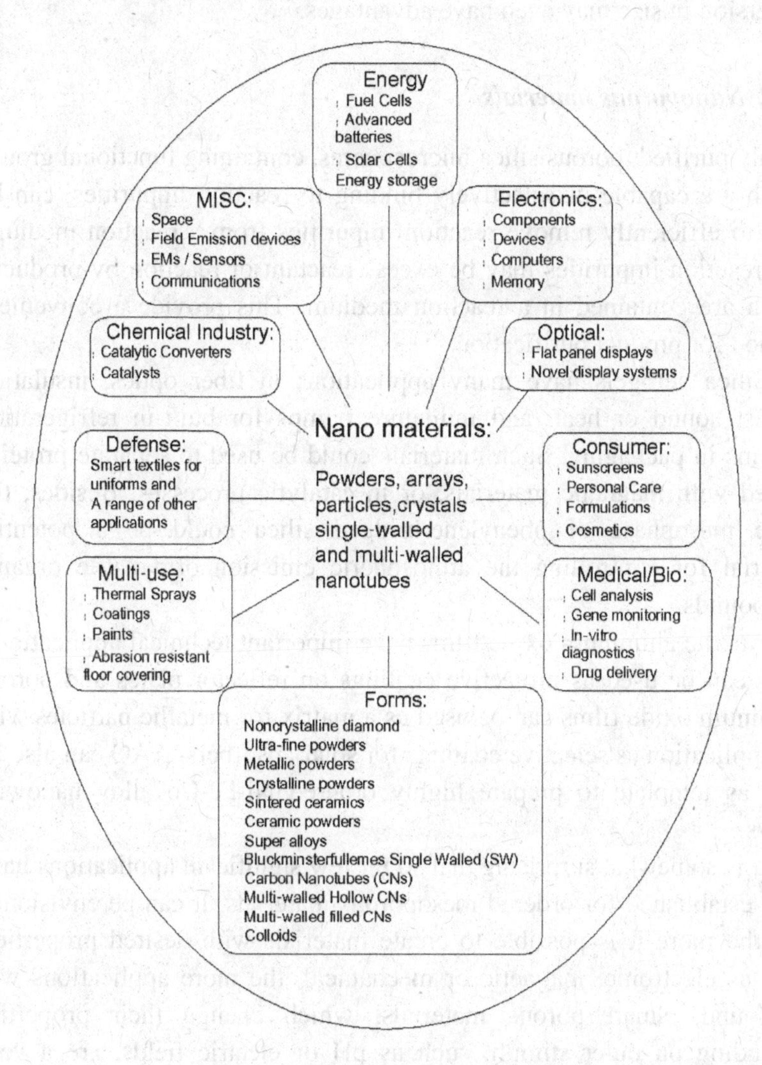

Figure 5.11 Applications of nanomaterials.

In many cases, such as iron oxide particles for magnetic tapes and nano silver bromide particles for photographic films, there is a clear industrial requirement for monodisperse nanoparticles. While in some applications, such as silica filler materials in the plastics industry, this requirement is not so strict. Sometimes, nanopowders with a controlled dispersion in size may even have advantages.

5.8.2 *Nanoporous materials*

Highly purified, porous silica microspheres, containing functional groups which are capable of selectively binding to reaction impurities, can be used to efficiently remove reaction impurities from a reaction medium. The reaction impurities may be excess reactant or reaction by-products, which are contained in a reaction medium. This provide a convenient method for product purification.

Silica aerogels have many applications in fiber optics, insulation against sound or heat, and miniature pumps for built-in refrigeration systems in packaging. Such materials could be used to separate proteins tagged with magnetic materials, or in catalytic processes. Besides, the cubic mesophase of phenylene-bridged silica could be a potential material for controlling the atmospheric emission of volatile organic compounds.

Anodic aluminum oxide films have important technical applications. They can be used as protective coatings on reflector plates and porous aluminum oxide films can be used as a matrix for metallic particles with the application as selective coatings for solar absorbers. AAO can also be used as template to prepare highly ordered Ni-Fe-Co alloy nanowire arrays.

It is somewhat surprising that so far few significant applications have been established for ordered mesoporous materials. It can be envisioned that the more it is possible to create materials with desired properties, such as electronic, magnetic or mechanical, the more applications will be found. Smart porous materials, which change their properties depending on outer stimuli, such as pH or electric fields, are a very interesting goal.

5.8.3 *Nanodusts*

Nanodusts could be any number of things, like particles of tungsten, silicon or photoresist material used in the manufacturing process. It could be something that flaked off from the side of the walls of the reaction chamber. But it has many things to offer in the world of materials. The biggest advantage it has is the low cost. The products made out of it are going to be very cost effective and it could provide altogether a new dimension to nanotechnology.

In the following, we discuss four typical applications: energy, pollution control, nuclear protection and composite reinforcement.

5.8.3.1 *Energy*

Presently, oil is the most important fuel and it is known to the world that this resource is limited. There is a need to think on many other possible sources of energy. Apart from many forms of energies, such as solar energy, electric energy, nuclear energy and hydro energy, the fuel of the future could be metal nanodusks.

Metals, such as iron, aluminum and boron, have tremendous promise. In nanodust form, the mixture of these metals are highly reactive. They can be ignited, and release high quantity of energy. With a modified engine and a tank full of metal nanodusts, an average saloon car could travel three times as far as the equivalent petrol-powered vehicle. Furthermore, this metal nano-fuel is almost completely nonpolluting because no carbon dioxide, no soot and no nitrogen oxides will be produced. In addition, it is rechargeable. By treating the spent nanodusts with a little hydrogen, the stuff can be burnt again. All kinds of engines, from domestic heating units to the turbines in power stations could be adapted to burn metal nanodusts.

However, iron nanodusts fuel has one obvious disadvantage. Though it is compact compared to hydrogen, iron nanodust fuel is very heavy. As the spent fuel is kept on board, unlike the polluting byproducts of the conventional fuels, this weight will not decrease. One solution to the weight problem is to use aluminum nanodusts rather than iron nanodusts and it gives four times as much energy per kilogram of iron. Boron nanodusts can provide six times as much. But these metals cost more than iron, and this would make the fuel more expensive.

5.8.3.2 *Pollution control*

Iron nanodusts are highly reactive, and they can readily combine with pollutant chemicals and break them down. They could dramatically reduce the levels of toxic chemicals at contaminated sites. Laboratory and field tests have shown that they can break down most common pollutants in several hours, such as trichloroethene, carbon tetrachloride and dioxins. They could break down a number of pollutants such as pesticides, organic dyes and chlorinated benzenes, albeit relatively slowly. It is expected that they can also be used to trap and break down industrial chemicals in ground water.

Under the right conditions, the metal breaks down water to produce hydrogen, as well as reacting with oxygen to form iron oxide, known as rust. But when other chemicals are present, the chemistry can be much more complex.

5.8.3.3 *Nuclear protection*

Four major effects take place when a nuclear explosion takes place: heat generation, sound effect, nuclear radiation and electromagnetic pulse (EMP). Among these effects, nuclear radiation has wide and long lasting effect. Moreover, the nuclear debris, with radiations of a few kms, can travel long distances with the wind, hence, causing wider effect. Here, nanodusts can play an important role. The nanodusts may be dispersed in atmosphere, such as stratosphere, so that they can absorb and scatter nuclear radiation, and reduce EMP effects by trapping free neutrons. Therefore, the adverse effects of a nuclear explosion can be minimized.

5.8.3.4 *Composite reinforcement*

Nanodusts are a promising reinforcing material which will produce cost effective composite structures for applications in many fields, such as aerospace, civil sectors and sports goods. Nanodusts, specially nanosilica, have a great potential in the development of composites where electromagnetic radiations are encountered, and avionics structures working at high temperatures.

5.8.4 *Nanowires*

Nanowires have shown great application promises. They may play important roles in electronic, opto-electronic and nanoelectromechanical devices. Nanowires can be used as field-emittors, interconnects in nanoscale quantum devices and leads for biomolecular nanosensors, and also they can be used as additives in advanced composites.

Among various types of nanowires, magnetic nanowires have been widely used for biomedical applications, such as drug delivery and cell manipulation. Due to their strong magnetization and strong mechanical strength, under external magnetic field, functionalized nanowires can go inside cells, and deliver drugs and genes there. This technique greatly improves the efficiency of normal drug and gene delivery. Cell manipulations mainly include cell separation, cell trapping and cell chaining, and are very important techniques in biomedical research. Here we discuss the applications of nanowires for cell chaining.

To chain cells, cell suspensions are placed in culture dishes. As shown in Figure 5.12, a uniform external field is applied to align the nanowires, and cell chains form under the effects of the external magnetic field. The nanowires experience mutually attractive dipole–dipole forces due to the interactions of their magnetic moments, and the alignment of the magnetic moments of nanowires favor the formation of head-to-tail chains, where the north pole of one wire abuts the south pole of the next. These formations can encompass many cells, and extend over hundreds of micrometers [35].

5.8.5 *Inorganic nanotubes*

The great diversity of inorganic nanotubes has considerably enlarged the possible applications already predicted for carbon nanotubes. In spite of their many similarities, especially in morphology and mechanical properties, pure inorganic nanotubes, or those co-grown with carbon nanotubes, show specific physical and chemical properties, based on

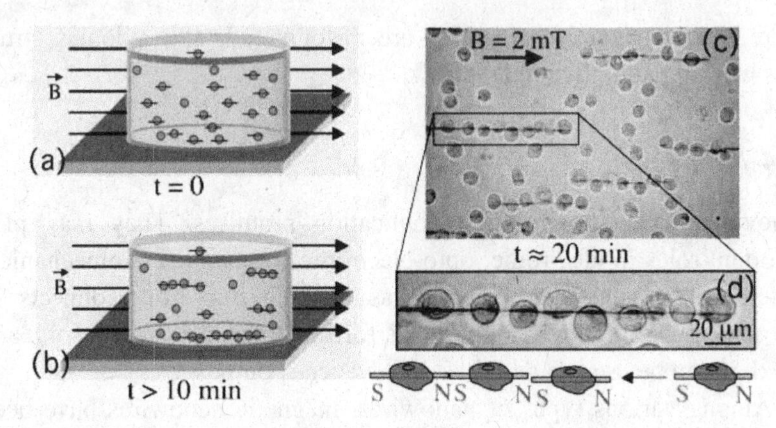

Figure 5.12 Magnetic cell chaining. (a) Schematic of nanowires bound to suspended cells and aligned in a magnetic field B. (b) Schematic of chain formation process due to magnetic dipole-dipole interactions between pre-aligned nanowires. (c) Cell chains formed on the bottom of a culture dish with B = 2 mT. (d) Close up of a single cell chain detailing wire-wire alignment. Interactions of north and south poles of adjacent wires are indicated schematically below [35]. Monica Tanase, Edward J. Felton, Darren S. Gray, Anne Hultgren, Christopher S. Chen and Daniel H. Reich, *Lab Chip*, 2005, **5**, 598-605, DOI: 10.1039/b500243e – *Reproduced by permission of The Royal Society of Chemistry.*

which a lot of new applications can be developed [20]. For example, due to their cylindrical geometry, these nanomaterials have a high porosity, a low mass density and an extremely large surface to weight ratio. Their potential applications range from high porous catalytic and ultralight anti-corrosive materials to electron field emitters and nontoxic reinforcement fibers.

Among various types of inorganic nanotubes, magnetic nanotubes (MNTs) exhibit many applications in biomedical research. Nanotubes are highly attractive for the development of multi-functional nanomaterials due to their structural features, such as the distinctive inner and outer surfaces, over the conventional nanopowders and nanowires. Because tube dimensions can be controlled in the synthesis procedure, the inner void of a nanotube can be used to capture, concentrate and release species ranging in size from large proteins to small molecules. The outer surface of a nanotube can be differentially functionalized with environment-friendly molecules and/or probe molecules to a specific target. By combining the structural features with magnetic properties,

MNTs are an ideal candidate for the multi-functional nanomaterial for biomedical applications, for instance, targeted drug delivery with MRI capability [36]. In the following, we discuss the applications of magnetic nanotubes in the control of bio-interactions and neuron growth.

5.8.5.1 *Control of bio-interactions*

The magnetic properties of MNTs can be used to facilitate and enhance the bio-interactions between the outer surfaces of MNTs and a specific target. For example, if the MNTs carrying drugs inside have probe molecules, such as anti-body, on the outer surfaces, the efficiency of drug delivery may be greatly improved by the magnetic-field-assisted bio-interaction.

To prove this concept, MNTs with 60 nm diameter are used. Their inner surfaces are modified with FITC, and their outer surfaces are modified with rabbit IgG. Such MNTs are added onto a glass slide modified with anti-rabbit IgG, and they are incubated for 10 min with and without a magnetic field from the bottom of the glass slide [36]. As shown in Figure 5.13, in the presence of the magnetic field, the

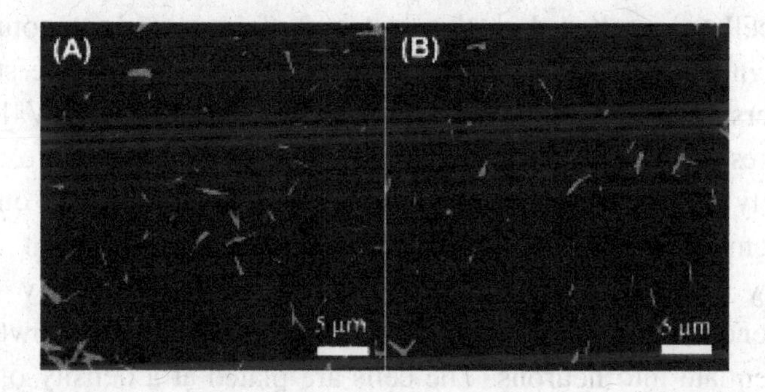

Figure 5.13 Fluorescence microscope image of bound MNTs, whose inner surfaces are FITC-modified and outer surfaces are rabbit IgG-modified, to anti-rabbit IgG-modified glass after antigen-antibody interaction with (A) and without (B) magnetic field under the glass substrate. The long bright spots in A and B are individual nanotubes, while some brighter long spots may be caused by the aggregation of two or three nanotubes [36]. Reprinted with permission from Son, S. J., Reichel, J., He, B., Schuchman, M. and Lee, S. B. (2005). Magnetic nanotubes for magnetic-field-assisted bioseparation, biointeraction, and drug delivery, *Journal of the American Chemical Society*, **127**, 7316-7317. © 2005 American Chemical Society.

antigen-antibody interactions are enhanced by about 4.2 fold. Therefore, the efficiency of antigen-antibody interactions can be controlled spatially by an external magnetic field.

5.8.5.2 *Growth of neuron cells*

Nervous system needs to function normally to enable animals to sense and respond to environmental stimuli by generating and conducting nerve impulses. Degeneration and injuries of nervous system result in severe disabilities and diseases, such as paralysis and Parkinson's disease. To overcome the challenges in the treatment of the disabilities and diseases of nervous systems, extensive efforts are underway to investigate and regulate the growth and activities of the basic structural and functional unit of a nervous tissue, namely the neuron, which is uniquely specialized to generate and conduct nerve signals. In recent years, much research attention is focused on the relationship between magnetic fields and neuron activities. Recent research indicates that magnetic nanotubes could be used in the control and regulation of neuron growth and activities.

To verify the biocompatibility of iron oxide (hematite) nanotubes, neuron cells are cultured in the presence of iron oxide nanotubes. At first, magnetic nanotubes with a diameter of 100-200 nm are sterilized by immersing in ethyl alcohol followed by exposure to UV light for 30 minutes, and then laid on sterile glass coverslips which were coated with poly D-lysine (60 μg/ml) and cells were plated on them. Pheochromocytoma cells (PC12 cells, derived from rat adrenal medullary tumor) are used, because these cells are widely used in neuroscience research and they need the presence of nerve growth factor to differentiate into neurons. The cells are plated at a density of 10,000 cells/dish on nanotube-laid glass coverslips kept in sterile 35 mm culture dishes. The differentiation of PC 12 cells are used to evaluate the cytotoxic effects, if any, of the nanotubes and the competence of the neurons to remain physiologically active. Once sufficient differentiation has occurred, the culture medium is collected for future lactate dehydrogenase assay (LDH) for biochemical determination of cell death. Then the dishes are fixed with 3% paraformaldehyde to be examined by light microscopy for noticeable differences in cell numbers and differentiation.

It is found that neuron cells grow well in the presence of iron oxide nanotubes, and some nanotubes are attached to the surface of neurites. As shown in Figure 5.14, neurites navigate across magnetic nanotubes laid on the glass substrate. At the same time, lots of nanotubes are found to locate on the surface of neurites implying a nonrepulsion, possibly nontoxic interaction between neurites and nanotubes. This work is helpful for us to understand the interaction between magnetic nanomaterials and neurons, and pave the way towards developing potential treatments using magnetic nanotubes for neurodegenerative disorders and injuries to the nervous system in the future [24].

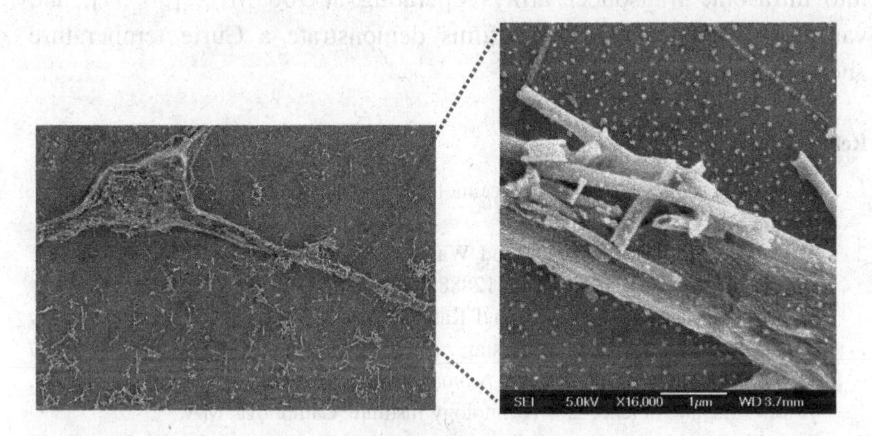

Figure 5.14 SEM micrographs of magnetic nanotubes attached to the surface of a neurite. (a) Low magnification; (b) higher magnification [24].

5.8.6 *ZnO nanostructures*

Due to their special properties, zinc oxide nanorods are attractive for photo-voltaic applications. Typical ZnO nanorods have a wurtzite structure with lattice spacing a = 0.32469 nm and c = 0.52069 nm, and they can be used for various applications in transparent electronics, ultraviolet light emitters, piezoelectric devices, chemical sensors and spin electronics [37-46].

Transparent thin film transistors (TFTs) with ZnO active channels have achieved much higher field effect mobility than amorphous silicon TFTs [47-49]. These transistors can be widely used for display applications. ZnO has been proposed to be a more promising UV emitting phosphor than GaN because of its larger exciton binding energy (60 meV). This leads to a reduced UV lasing threshold and yields higher UV emitting efficiency at room temperature [50]. Surface acoustic wave filters using ZnO films have already been used for video and radio frequency circuits. Hole mediated ferromagnetic ordering in bulk ZnO by introducing Mn as dopant has been predicted theoretically [51]. Bulks and thin films of ZnO have demonstrated high sensitivity for toxic gases [52-55]. Furthermore, piezoelectric ZnO thin films have been fabricated into ultrasonic transducer arrays operating at 100 MHz [56, 57], and vanadium doped n-type ZnO films demonstrate a Curie temperature above room temperature [58].

References:

1. Whatmore, R. W. (2001). Nanotechnology: Big prospects for small engineering, *Ingenia*, 28-33 (February).
2. Kirkland, J. J., Langlois, T. J. and Wang, Q. (2006). *Porous Silica Microsphere Scavengers*, United States Patent 7128884.
3. Rissanen, A. K., Blomberg, M. and Kattelus, H. (2004). Improved reproducibility in porous silica sol-gel processing using tertiary butanol solvent, *Technical Proceedings of the 2004 NSTI Nanotechnology Conference and Trade Show*, Volume 3, Nano Science and Technology Institute, Cambridge, MA.
4. Ono, M., Yoshii, K., Kuroki, T., Yamasaki, N., Tsunematsu, S. and Bignall, G. (2001). Development of porous silica production by hydrothermal method, *High Pressure Research*, **20**, 307-310.
5. Takeuchi, Y. (1999). *Production of Porous Materials, Porous Materials*, Fuji Technosystem, Japan.
6. Kimura, T., Sugahara, Y. and Kuroda, K. (1999). Synthesis and characterization of lamellar and hexagonal mesostructured aluminophosphates using alkyltrimethyl-ammonium cations as structure-directing agents, *Chemistry of Materials*, **11**, 508-518.
7. Furneaux, R. C. and Rigby, W. R. (1989). *Porous Anodic Aluminum Oxide Films*, United States Patent 4859288.
8. Lagrene, K. and Zanotti, J. M. (2007). Anodic aluminium oxide: Concurrent SEM and SANS characterisation. Influence of AAO confinement on PEO mean-square displacement, *The European Physical Journal - Special Topics*, **141**, 261-265.

9. Davies, N. C. and Sheasby, P.G. (1987). *Anodic Aluminum Oxide Film and Method of Forming It*, United States Patent 4681668.
10. Imamutdinov, I. (2003). Ground to Nano-dust, *Gateway to Russia*, *www.gateway2russia.com*.
11. Catchpole, H. (2004). Nanodust sends tornadoes into a spin, *ABC Science Online*, *www.abc.net.au/science*.
12. Amans, D., Callard, S., Gagnaire, A. and Joseph, J., Huisken, F. and Ledoux, G. (2004). Spectral and spatial narrowing of the emission of silicon nanocrystals in a microcavity, *Journal of Applied Physics*, **95** (9), 5010-5013.
13. Habbal, S. R., Arndt, M. B., Nayfeh, M. H., Arnaud, J., Johnson, J., Hegwer, S., Woo, R., Ene, A. and Habbal, F. (2003). On the detection of the signature of silicon nanoparticle dust grains in coronal holes, *Astrophysical Journal*, **592**, L87-L90.
14. Nayfeh, M. H., Barry, N., Therrien, J., Akcakir, O., Gratton, E., Belomoin, G. (2001). Stimulated blue emission in reconstituted films of ultrasmall silicon nanoparticles, *Applied Physics Letters*, **78** (8), 1131-1133.
15. Mead, T. (2003). Iron nano dust zaps tricky pollutants, *New Scientist*, **179** (2413), 22-22.
16. Bauer, L. A., Birenbaum, N. S. and Meyer, G. J. (2004). Biological applications of high aspect ratio nanoparticles, *Journal of Materials Chemistry*, **14**, 517-526.
17. Sun, L., Hao, Y., Chien, C. L. and Searson, P. C. (2005). Tuning the properties of magnetic nanowires, *IBM Journal of Research and Development*, **49** (1), 79-102.
18. Bauer, L. A., Reich, D. H. and Meyer, G. J. (2003). Selective functionalization of two-component magnetic nanowires, *Langmuir*, **19**, 7043-7048.
19. Bera, D., Kuiry, S. C. and Seal, S. (2004). Synthesis of nanostructured materials using template-assisted electrodeposition, *JOM*, **56**(1), 49–53.
20. Remskar, M. (2004). Inorganic nanotubes, *Advanced Materials*, **16**, 1497-1504.
21. Patzke, G. R., Krumeich, F. and Nesper, R. (2002). Oxidic nanotubes and nanorods - anisotropic modules for a future nanotechnology, *Angewandte Chemis – International Edition*, **41**, 2446-2461.
22. Tenne, R., Margulis, L., Genut, M. and Hodes, G. (1992). Polyhedral and cylindrical structures of tungsten disulfide, *Nature*, **360**, 444-446.
23. Sui, Y. C., Skomski, R., Sorge, K. D. and Sellmyer, D. J. (2004). Magnetic nanotubes produced by hydrogen reduction, *Journal of Applied Physics*, **95**, 7151-7153.
24. Chen, L. F., Xie, J. N., Yancey, J., Srivatsan, M. and Varadan, V. K. (2007). Experimental investigation of magnetic nanotubes in PC-12 neuron cells culturing, *Proceedings of the SPIE*, **6528**, 65280L.
25. Fujita, S., Kim, S. W., Ueda, M. and Fujita, S. (2004). Artificial control of ZnO nanostructures grown by metalorganic chemical vapor deposition, *Journal of Crystal Growth*, **272**, 138-142.

26. Zhou, H. L. and Li, Z. (2005). Synthesis of nanowires, nanorods and nanoparticles of ZnO through modulating the ratio of water to methanol by using a mild and simple solution method, *Materials Chemistry and Physics*, **89**, 326-331.
27. Zeng, D. W., Xie, C. S., Dong, M., Jiang, R., Chen, X., Wang, A. H., Wang, J. B. and Shi, J. (2004). Spinel-type $ZnSb_2O_4$ nanowires and nanobelts synthesized by an indirect thermal evaporation, *Applied Physics A – Materials Science & Processing*, **79**, 1865-1868.
28. Zhang, J., Sun, L. D., Jiang, X. C., Lian, C. S. and Yan, C. H. (2004). Shape evolution of one-dimensional single-crystalline ZnO nanostructures in a micro-emulsion system, *Crystal Growth and Design*, **4**, 309-313.
29. Zhang, H., Yang, D., Ma, X., Ji, Y., Xu, J. and Que, D. (2004). Synthesis of flower-like ZnO nanostructures by an organic-free hydrothermal process, *Nanotechnology* **15**, 622-626.
30. Gao, P. X. and Wang, Z. L. (2004). Substrate atomic-termination-induced anisotropic growth of ZnO nanowires/nanorods by the VLS process, *Journal of Physical Chemistry B*, **108**, 7534-7537.
31. Leung, Y. H., Djurišić, A. B., Gao, J., Xie, M. H., Wei, Z. F., Xu, S. J. and Chan, W. K. (2004). Zinc oxide ribbon and comb structures: synthesis and optical properties, *Chemical Physical Letters*, **394**, 452-457.
32. Liu, F., Cao, P. J., Zhang, H. R., Li, J. Q. and Gao, H. J. (2004). Controlled self-assembled nanoaeroplanes, nanocombs, and tetrapod-like networks of zinc oxide, *Nanotechnology,* **15**, 949-952.
33. Hu, P. ALiu, ., Y. Q., Fu, L., Wang, X. B. and Zhu, D. B. (2004). Creation of novel ZnO nanostructures: self-assembled nanoribbon/nanoneedle junction networks and faceted nanoneedles on hexagonal microcrystals, *Applied Physics A – Materials Science and Processing*, **78**, 15-19.
34. Tian, Z. R., Voigt, J. A., Liu, J., Mckenzie, B. and Mcdermott, M. J. (2003). Complex and oriented ZnO nanostructures, *Nature Materials*, **2**, 821-826.
35. Tanase, M., Felton, E. J., Gray, S., Hultgren, A., Chen, C. S. and Reich, D. H. (2005). Assembly of multicellular constructs and microarrays of cells using magnetic nanowires, *Lab on a Chip*, **5**, 598-605.
36. Son, S. J., Reichel, J., He, B., Schuchman, M. and Lee, S. B. (2005). Magnetic nanotubes for magnetic-field-assisted bioseparation, biointeraction, and drug delivery, *Journal of the American Chemical Society*, **127**, 7316-7317.
37. Nomura, K., Ohta, H., Ueda, K., Kamiya, T., Hirano, M. and Hosono, H. (2003). Thin-film transistor fabricated in single-crystalline transparent oxide semiconductor, *Science*, **300**, 1269-1272.
38. Nakada, T., Hirabayashi, Y., Tokado, T., Ohmori, D. and Mise, T. (2004). Novel device structure for Cu(In,Ga)Se-2 thin film solar cells using transparent conducting oxide back and front contacts, *Solar Energy*, **77**, 739-747.
39. Lee, S. Y., Shim, E. S., Kang, H. S., Pang, S. S. and Kang, J. S. (2005). Fabrication of ZnO thin film diode using laser annealing, *Thin Solid Films*, **437**, 31-34.

40. Könenkamp, R., Word, R. C. and Schlegel, C. (2004). Vertical nanowire light-emitting diode, *Applied Physics Letters*, **85**, 6004-6006.
41. Trolier-McKinstry, S. and Muralt, P. (2004). Thin film piezoelectrics for MEMS, *Journal of Electroceramics*, **12**, 7-17.
42. Wang, Z. L., Kong, X. Y., Ding, Y., Gao, P., Hughes, W. L., Yang, R. and Zhang, Y. (2004). Semiconducting and piezoelectric oxide nanostructures induced by polar surfaces, *Advanced Functional Materials*, **14**, 943-956.
43. Wagh, M. S., Patil, L. A., Seth, T. and Amalnerkar, D. P. (2004). Surface cupricated SnO_2-ZnO thick films as a H_2S gas sensor, *Materials Chemistry and Physics*, **84**, 228-233.
44. Ushio, Y., Miyayama, M. and Yanagida, H. (1994). Effects of interface states on gas-sensing properties of a CuO-ZnO thin film heterojuntion, *Sensors and Actuators B – Chemical*, **17**, 221 (1994).
45. Harima, H. (2004). Raman studies on spintronics materials based on wide bandgap semiconductors, *Journal of Physics – Condensed Matter*, **16**, S5653-S5660.
46. Pearton, S. J., Heo, W. H., Ivill, M., Norton, D. P. and Steiner, T. (2004). Dilute magnetic semiconducting oxides, *Semiconductor Science and Technology*, **19**, R59-R74.
47. Nishii, J., Hossain, F.M., Takagi, S., Aita, T., Saikusa, K., Ohmaki, Y., Ohkubo, I., Kishimoto, S., Ohtomo, A., Fukumura, T., Matsukura, F., Ohno, Y., Koinuma, H., Ohno, H. and Kawasaki, M. (2003). High mobility thin film transistors with transparent ZnO channels, *Japanese Journal of Applied Physics Part 2 – Letters*, **42**, L347-L349.
48. Hossain, F. M., Nishii, J., Takagi, S., Sugihara, T., Ohtomo, A., Fukumura, T., Koinuma, H., Ohno, H. and Kawasaki, M. (2004). Modeling of grain boundary barrier modulation in ZnO invisible thin film transistors, *Physica E – Low Dimensional Systems & Nanostructures*, **21**, 911-915.
49. Norris, B. J., Anderson, J., Wager, J. F. and Kszler, D. A. (2003). Spin-coated zinc oxide transparent transistors, *Journal of Physics D – Applied Physics*, **36**, L105-L107.
50. Yang, P., Yan, H., Mao, S., Russo, R., Johnson, J., Saykally, R., Morris, N., Pham, J., He, R. and Choi, H. J. (2002). Controlled growth of ZnO nanowires and their optical properties, *Advanced Functional Materials*, **12**, 323-331.
51. Dietl, T. (2002). Ferromagnetic semiconductors, *Semiconductor Science and Technology,* **17**, 377-392.
52. Ryu, H. W., Park, B. S., Akbar, S. A., Lee, W. S. Hong, K. J., Seo, Y. J, Shin, D. C., Park, J. S. and Choi, G. P. (2003). ZnO sol-gel derived porous film for CO gas sensing, *Sensors and Actuators B – Chemical*, **96**, 717-722.
53. Sberveglieri, G. (1995). Recent developments in semiconducting thin-film gas sensors, *Sensors and Actuators B – Chemical*, **23**, 103-109.
54. Rao, G. S. T. and Rao, D. T. (1999). Gas sensitivity of ZnO based thick film sensor to NH3 at room temperature, *Sensors and Actuators B – Chemical*, **55**, 166-169.

55. Cheng, X. L., Zhao, H., Huo, L. H., Gao, S. and Zhao, J. G. (2004). ZnO nanoparticulate thin film: preparation, characterization and gas-sensing property, *Sensors and Actuators B – Chemical*, **102**, 248-252.
56. Ito, Y., Kushida, K., Sugawara, K. and Takeuchi, H. (1995). A 100-MHz ultrasonic transducer array using ZnO thin films, *IEEE Transactions on Ferroelectrics, and Frequency Control*, **42**, 316-324.
57. Sharma, P., Gupta, A., Rao, K. V., Owens, F. J., Sharma, R., Ahuja, R., Osorio, J. M., Johansson, B. and Gehring, G. A. (2003). Ferromagnetism above room temperature in bulk and transparent thin films of Mn-doped ZnO, *Nature Materials,* **2**, 673-677.
58. Saeki, H., Tabata, H. and Kawai, T. (2001). Magnetic and electric properties of vanadium doped ZnO films, *Solid State Communications*, **120**, 439-443.

Chapter 6

Polymer Nanotechnology

Polymer nanostructures and microstructures are playing important roles in many fields, especially medical engineering. After introducing electroactive polymers, this chapter discusses the fabrication of polymer nanowires, polymer nanotubes and three-dimensional (3D) polymer microstructures.

6.1 Electroactive Polymers

Electroactive polymers have been widely used for the development of smart materials and structures. In the following subsections, we discuss two typical types of electroactive polymers: polypyrrole and Nafion.

6.1.1 *Polypyrrole*

Electrically conductive polymers, such as polypyrrole (PPy), have been extensively studied because of their unique properties of high conductivity and high chemical stability, and their ease for preparation. These materials promise new applications in many technological areas, including electroactive polymer actuators, batteries, electrochemical sensors and functional electronic devices. Polypyrrole is made up of pyrrole monomers, C_4H_5N which is an organic compound with a ring structure composed of four carbon atoms and one nitrogen atom. Despite the availability of many synthesis methods, electropolymerization is among the most popular techniques for preparing PPy films on electrodes. Since it can take place in aqueous solutions of pyrrole allowing a variety of dopant possibilities, the morphology and electrochemical properties of polypyrrole films can be adjusted by various dopant ions. One of the dopants is Dodecylbenzene Sulfonate (DBS-), a large bulky anion. It is known that DBS- ions can produce polypyrrole films with high conductivity and great mechanical transduction properties. Understanding how dopant anions interact within the film and change the structural and physical properties as well as the stability and reactivity of the polypyrrole are particularly important for developing new and improving existing technological applications of polypyrrole.

6.1.2 *Nafion*

Nafion is a Teflon-based polymer with sulfonic acid side groups and, can be used as an ion exchange polymer for applications of battery, ionic polymer transducing and ion selective sensing devices. When nafion is hydrated in water, it can exchange cations in the polymer membrane. The cations electrostatically attracted to the sulfonate group become mobile and leave sulfonate anions which is covalently bonded to the fluorocarbon backbone. It allows ion exchange polymers to exhibit electromechanical transduction, also to serve as an ion selective membrane (ISM) for cation sensing. In ionic polymer actuating applications, ions move in and out of the conducting polymers under an electric field, which lead to simultaneous changes in volume as well as variations in physical properties. Vice versa, if a thin strip of Nafion is subjected to a bending moment with lateral force which produces a stress on the backbone polymer, an electric potential is generated across the composite. It is assumed that the potential generation is due to the differential displacement of the effective centers of anions and cations within each cluster, producing an effective dipole [1].

6.2 Fabrication of Polymer Nanowires

As polymer nanostructures have low density and high surface density, and their surface properties can be easily modified, extensive research has been made on the fabrication of polymer nanostructures over the past decades. Among many fabrication methods, electrochemical polymerization is introduced in this section, followed by critical point drying technique.

6.2.1 *Electrochemical polymerization*

Electrochemical deposition uses an oxidative process applying positive potential on the target surface in monomer electrolyte solution. This method is convenient to set up and can be applied to various surface geometries. Various factors influence the thickness and morphology of the deposited film such as oxidation potential and ionic dopant types. For example, polypyrrole nanowires can be fabricated inside a nanopore membrane which is attached on an anode surface. In a pyrole monomer solution containing 0.1 M NaClO4 and 0.05 M pyrrole, a positive potential is applied to electropolymerize the pyrrole inside the nanopores on the anode surface. Figure 6.1 shows an SEM image of polypyrrole nanowires grown on a substrate [2].

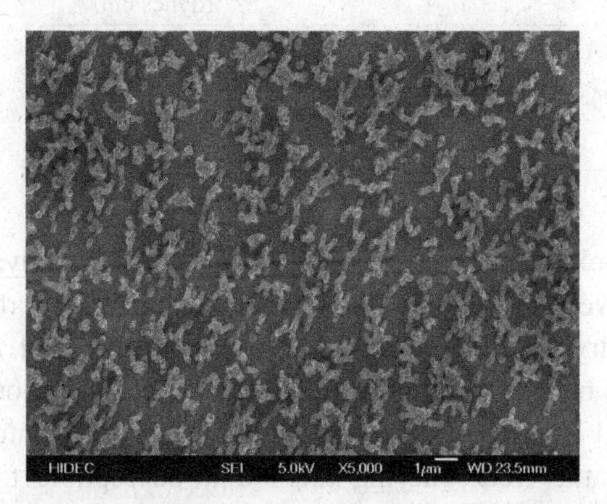

Figure 6.1 SEM image of polypyrrole nanowires grown on a substrate [2]. Reprinted with permission from Dubey, R., Shami, T. C., Rao, K. U. B., Yoon. H. and Varadan, V. K. (2009). Synthesis of polyamide microcapsules and effect of critical point drying on physical aspect, *Smart Materials and Structures*, **18** (2), 025021. IOP Publishing Ltd.

6.2.2 *Critical point drying*

In processing polymer nano- and microstructures, these polymer structures are apt to become defective because of the harsh environment during wet chemical cleaning and drying. One of the critical failures is the deformation during the drying process. It is believed that strong capillary force is the primary cause of such deformation and collapse of structure, due to the effect of surface tension. The most common medium for wet chemical process is water which has very high surface tension to air. While drying, a highly flexible polymer structure is subject to significant surface tension present at the phase boundary as liquid evaporates from the surface or inside the structure. By using a liquid with a lower surface tension, such as alcohol, as shown in Table 6.1, the surface tension can be decreased. However, it is not enough to prevent structural collapse for structures made of weak mechanical strength material, especially polymer.

Table 6.1 Surface tension of various substances at 20°C.

Substance	**Surface tension** (dynes/cm)
Ethanol	22.3
Acetone	23.7
Isopropanol	21.7
Glycerol	63.0
Water	72.8

Among polymeric micro/nanostructures, hollow polyamide microcapsules and vertically aligned nanowires have been reported to be very sensitive to drying conditions [2, 3]. Spray drying and freeze drying methods have been widely used; however, their applications are limited by mechanical rigidity of polymer structure and temperature condition. For example, the freeze drying method is susceptible to cracking of the polymer shell [4]. Last decades, critical point drying (CPD) process has been developed and widely used for drying of biological specimens [30] (see Figure 6.2) and releasing of moving structures in microelectromechanical systems (MEMS), because of the high yield and the simplicity of this process. At the critical point, the densities of the liquid and gas phases are equal and beyond this, phase boundary of the two phases disappears. Therefore, this homogeneous supercritical fluid can dry microcapsules without facing the issue of surface tension. In many applications, CO_2 is selected as the medium of critical point drying, because of its critical point near room temperature (31°C and 1,070 psi) compared to other material (see Table 6.2).

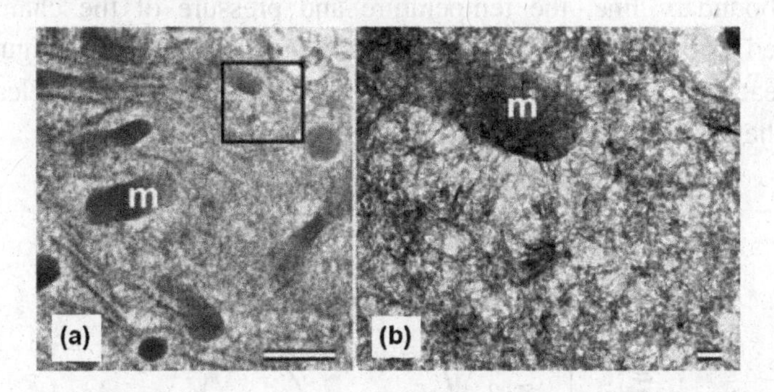

Figure 6.2 Ultrastructure under 100 kV SEM of rat liver section cut with thickness (<1 μm) as thinly as possible on cryostat and processed by critical point drying. m: mitochondria.. Area enclosed by rectangle in (a) is shown at higher magnification in (b). Bars: 1 μm in (a), and 0.1 μm in (b). Kondo, H. (2008). What we have learned and will learn from cell ultrastructure in embedment-free section electron microscopy, *Microscopy Research and Technique*, **71**, 418–442. © 2008. Reprinted with permission from John Wiley and Sons Inc.

Table 6.2 Temperature and pressure condition to reach critical point.

Substance	**Temperature** (°C)	**Pressure** (psi)
Argon	−122	707
Oxygen	−119	732
Nitrogen	−147	492
Carbon dioxide	+31	1,070
Carbon monoxide	−141	528
Water	+374	3,200

For critical point drying after wet chemical process, water should be exchanged with an intermediate dehydrate fluid such as isopropanol (IPA), because water is not miscible with CO_2. Usually, the specimen with polymer structures in water are processed gradually, through varying IPA concentrations (10%, 30%, 70%, 90% and 100%), thus completely replacing the water with the IPA before critical point drying. In dry processing, suspended microcapsules in IPA are loaded in the CPD chamber lid, the chamber temperature is cooled down to 8°C and liquid

CO_2 is loaded to fill the chamber. By flushing of liquid CO_2 and filtering IPA for 20 min, IPA in the chamber is replaced with CO_2 liquid. In order to pass around the CO_2 supercritical point without crossing over the liquid-air phase boundary line, the temperature and pressure of the chamber is increased to 35°C and 1,350 psi with the order as shown in Figure 6.3. After reaching the supercritical point, the chamber pressure is released to atmospheric pressure at 35°C by passing the phase boundary line.

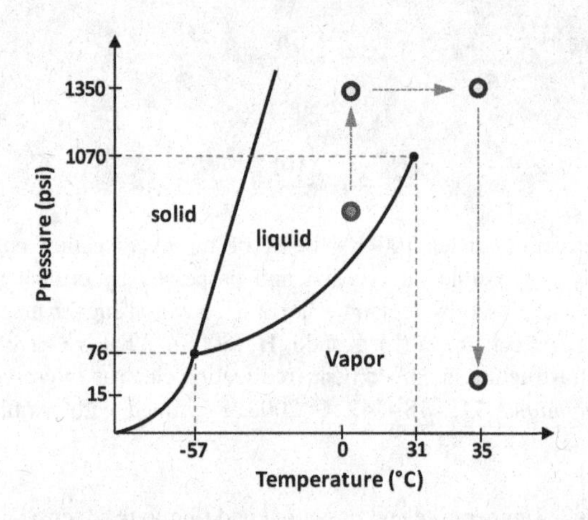

Figure 6.3 Schematic diagram of CO_2 critical point drying method.

(a) (b)

Figure 6.4 SEM images of (a) clumped polypyrrole nanowires after drying in the air and (b) a magnified side view image [2]. Reprinted with permission from Dubey, R., Shami, T. C., Rao, K. U. B., Yoon. H. and Varadan, V. K. (2009). Synthesis of polyamide microcapsules and effect of critical point drying on physical aspect, *Smart Materials and Structures*, **18** (2), 025021. IOP Publishing Ltd.

Figure 6.4 shows polypyrrole nanowires dried in the air. The clumped nanowires due to high surface tension of water during the drying process can be observed.

6.3 Fabrication of Polymer Nanotubes

The discovery of carbon nanotubes has stimulated a worldwide investigation of various kinds of nanotubes [5]. Besides carbon nanotubes, both ceramic nanotubes, such as SiO_2, TiO_2, SnO_2 and Fe_3O_4, and organic nanotubes, such as PMMA, PAN, Polystyrene and PVDF, have attracted considerable research interests. Among them, conducting polymer nanotubules, such as polyaniline (PANi), poly(2-methoxyaniline), poly(3-methylthiophene) and polypyrrole (PPy), are extremely attractive due to their desirable electrical, electrochemical and optical properties coupled with excellent environmental stability. It has been demonstrated that conducting polymer nanotubules can have an order of magnitude higher conductance than bulk materials [6].

Template synthesis is an elegant approach to fabricate nanomaterials. It has a number of unique features. Anodic aluminum oxide (AAO) and polycarbonate (PC) are two types of templates which are commonly used. Both of them have an array of cylindrical holes with a diameter in the range of 4 to 400 nm. The dimension of templates depends upon the preparation method. For an AAO template, the pore size and length of the nanopores are determined by the reaction condition of aluminum anodic oxidation.

Martin *et al.* has reported their preparation of nanofibrils composed of metals, semiconductors and conducting polymers from commercially available template membranes [7-10]. A number of research groups have worked on electropolymerization of conducting polymer nanotubules with template methods [11-13].

Encapsulation is often required to protect metal nanowires against corrosive environment because of their extremely high reactivities due to their nanoscale dimensions. For example, iron nanowires can be easily corroded in humid air. Open-ended conducting polymer nanotubules can be used as the second time-template to prepare metal nanowires inside nanotubules. Cao *et al*. reported an array of iron nanowires encapsulated in polyaniline nanotubules [14].

Metal nanowires encapsulated by conducting polymer nanotubes could find potential applications in electromagnetic wave absorber, field emission devices, perpendicular magnetic recording materials and so on. In the following, we discuss the synthesis of blank polymer nanotubes and nickel nanowires encapsulated by polyaniline nanotubues using electropolymerization/template method [15].

6.3.1 *Blank polymer nanotubes*

The materials used mainly include aniline, ACS reagent >99.5% (Aldrich), ACS reagent of hydrochloric acid 37% (Aldrich), and anodic alumina membranes with a nominal pore diameter of 200 nm and a thickness of 60 μm (VWR).

Before solution preparation, aniline needs to be distilled under vacuum to further purify the monomer. High Q water with 18.3 MΩ-cm of resistivity is used to prepare all the solutions. 0.3 M aniline in 1.0 M hydrochloric acid aqueous solution is prepared for electropolymerization of polyaniline nanotubules.

Before electropolymerization, electrode needs to be deposited on AAO membrane. First, a 300 nm thick silver thin film was deposited on the back side of AAO membrane by sputtering. In the second step, another silver film was deposited on the backside of membrane by electrochemical plating to cover all the area of membrane backside. This silver film also functioned as the electrode for further electrodeposition steps. After this step, silver electrodeposition was performed again on

Blank membrane (Anodic alumina)

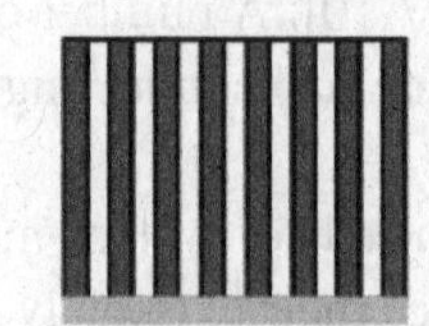

Backside contact formation (Ag by sputtering and electrodeposition)

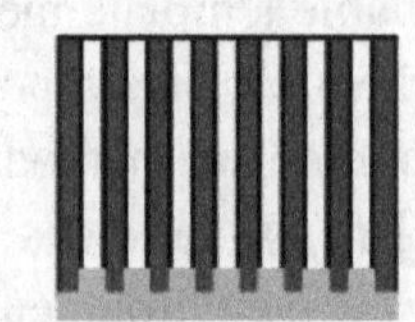

Front side electrodeposition (Ag)

Figure 6.5 Schematic diagram of electrode deposition on AAO membrane [15].

the front side of membrane to fill the possible gaps between membrane and the silver film deposited in the first two steps. Figure 6.5 shows a schematic diagram of electrode preparation on AAO template membranes.

A single electrochemical cell was used for polyaniline electropolymerization, as shown in Figure 6.6. The prepared aniline aqueous solution was pored into the cell, and then a Pt wire was immersed as the counter electrode. The silver coated AAO membrane was placed in such a way that monomer solution could access to the electrode only through the nanopores in the membrane. The silver thin film was connected to the anode of a DC power supply. During the electropolymerization, the electropolymerization current was monitored by a multimeter. Electropolymerization occurred at +0.8 V versus calomel reference electrode.

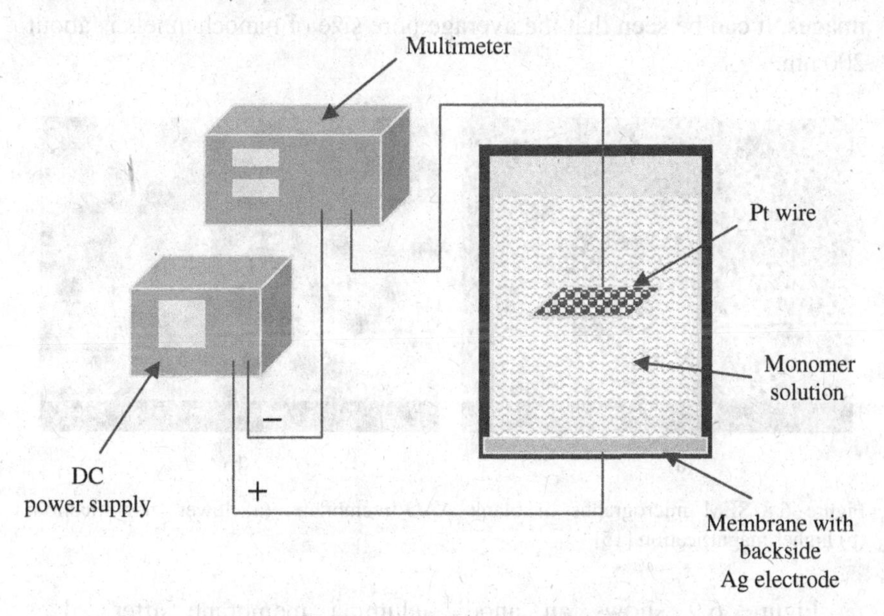

Figure 6.6 Schematic diagram of the electropolymerization setup [15].

After electropolymerization, polyaniline nanotubules formed inside the nanopores of AAO membrane, as shown in Figure 6.7. To release polyaniline nanotubules, AAO membrane was removed by immersing into a 3 M NaOH aqueous solution.

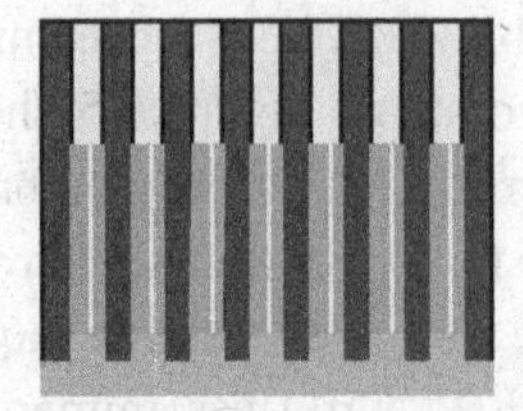

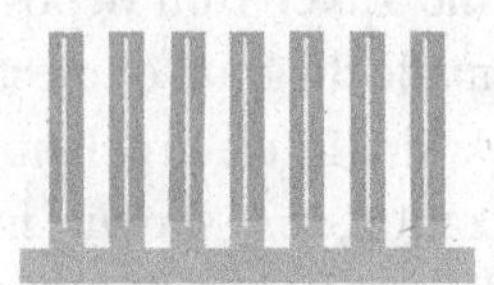

Figure 6.7 Schematic diagram of electropolymerization of polyaniline and template removing [15].

In the template method, the diameter of nanofibrils depends upon the pore size of nanochannels in the template. Figure 6.8 shows SEM micrographs of a blank AAO membrane before deposition. From SEM images, it can be seen that the average pore size of nanochannels is about 200 nm.

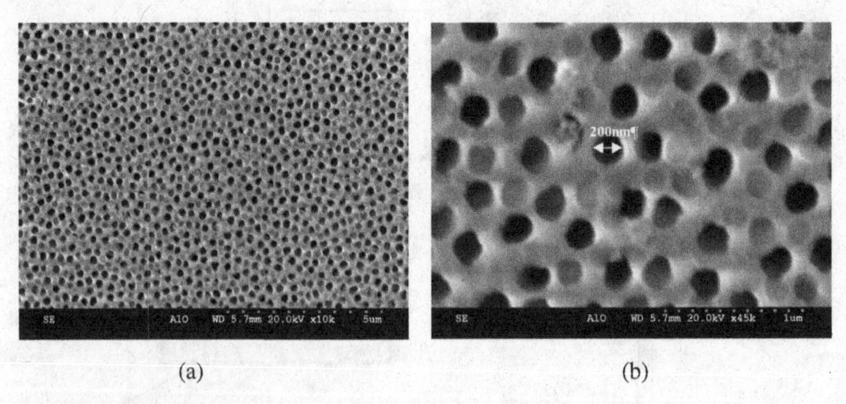

Figure 6.8 SEM micrographs of blank AAO membrane: (a) lower magnification; (b) higher magnification [15].

Figure 6.9 shows an anodic alumina membrane after silver deposition from backside. It is obvious that all of the nanochannel entrances are partially covered by a layer of silver. That is the reason why silver electrodeposition from front side is necessary. After all silver depositions are done, membranes are completely covered.

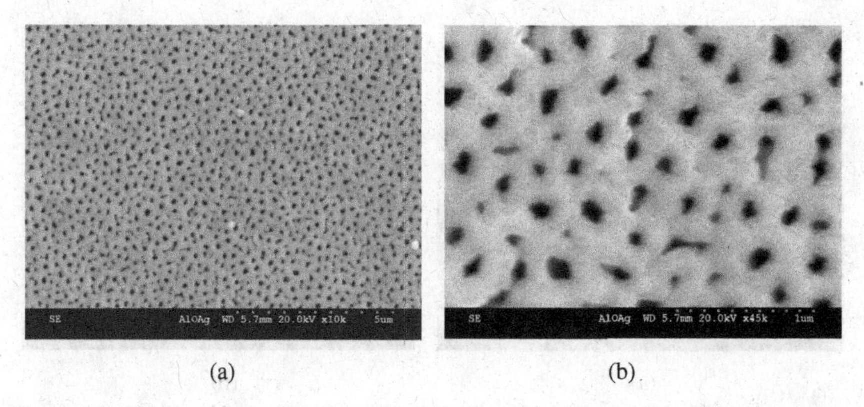

Figure 6.9 SEM micrographs of membrane after first silver electrodeposition from backside [15]: (a) lower magnification; (b) higher magnification.

During polyaniline electropolymerization, bubbles were observed from platinum wires indicating the start of reaction. The electro-polymerization current was monitored by a multimeter. Figure 6.10 shows the chronoamperogram of polyaniline electrochemical deposition for synthesizing polyaniline nanotubes. The electropolymerization current dropped at the beginning and rise up after about 10 mins. This could be explained by the diffusion time of monomer solution into the nanochannels.

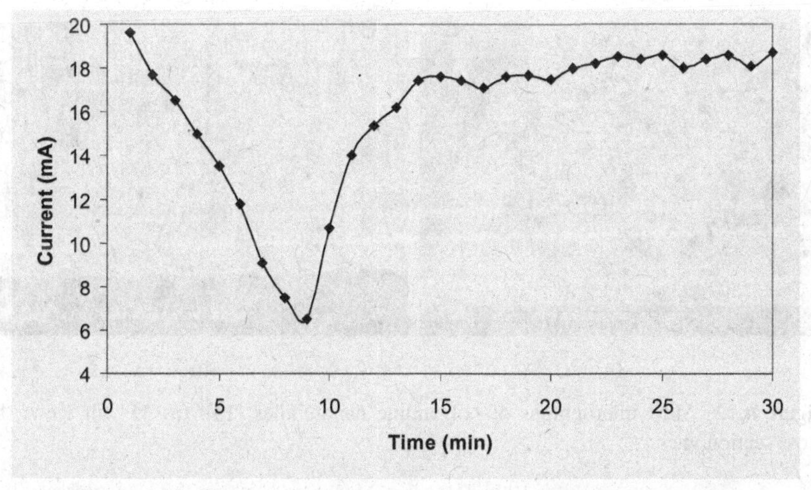

Figure 6.10 Chronoamperogram of polyaniline electropolymerization.

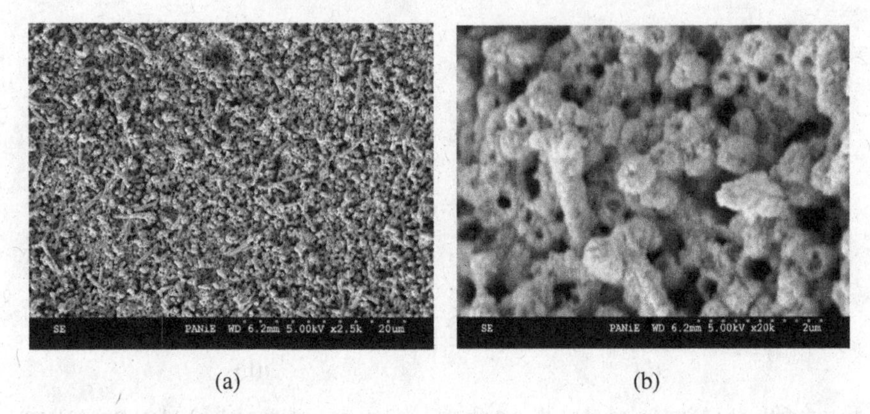

(a) (b)

Figure 6.11 SEM micrographs of polyaniline nanotubules after template etching [15]: (a) lower magnification; (b) higher magnification.

In a typical reaction, polyaniline electropolymerization is set to last for about one hour. Figures 6.11 (a) and (b) are SEM micrographs of the released polyaniline nanotubules after NaOH etching. Aligned nanotubules with approximately 200 nm in diameter were observed clearly. These nanotubules exhibit rough surface morphology, which is due to the rapid polymerization rate. Nanotubes with smooth surfaces can be obtained by adjusting electrolyte concentration and polymerization current.

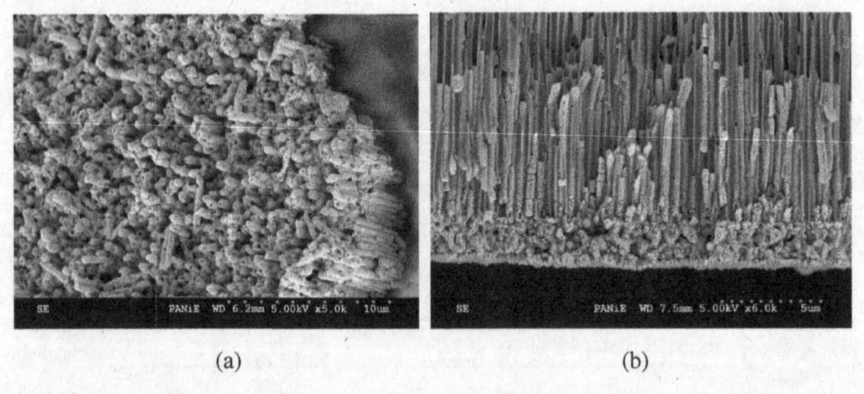

(a) (b)

Figure 6.12 SEM micrographs of polyaniline nanotubules [15]: (a) 45° tilt view; (b) cross section view.

Figure 6.12 (a) is an SEM image of 45° tilted polyaniline nanotubules. Figure 6.12 (b) shows cross section view of polyaniline nanotubules inside the AAO template. This image reveals grown polyaniline nanotubules inside AAO nanochannels. Silver thin film and silver particles were also observed at the backside of AAO template, serving as electrode for electropolymerization. The average length of the nanotubules is around 7 microns. One of the advantages of the electropolymerization method over a chemical method is that it is easy to control the growth rate as well as the nanotubule length. Basically, longer electropolymerization time leads to longer polyaniline nanotubules.

6.3.2 *Metal nanowires encapsulated in polymer nanotubes*

To synthesize metal nanowires encapsulated in polymer nanotubes, the polyaniline nanotubule array before template etching can be used as a "second" template. Figure 6.13 depicts the procedure for nickel encapsulation. In order to make sure nickel deposition occurs near the silver electrode, conducting polyaniline nanotubules need to convert to emeraldine base (EB) form, which is nonconducting. Therefore, blank polyaniline nanotubules inside AAO membrane were treated by diluted NH_4OH. Then regular nickel electroplating was performed to deposit nickel inside the polyaniline nanotubules. Again, nickel nanowires encapsulated in polyaniline nanotubules can be released by template etching. In order to re-obtain their conductivity, polyaniline nanotubules need to be chemically treated by diluted HCl solution.

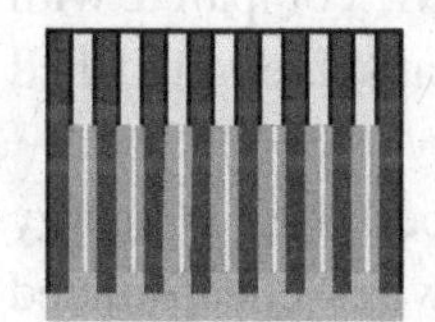

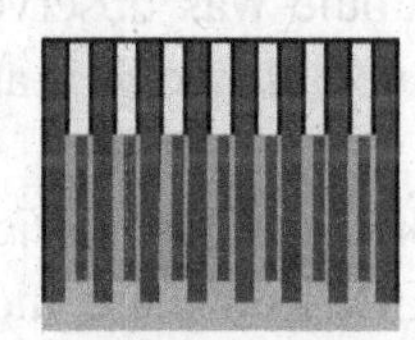

Figure 6.13 Schematic diagram of nickel encapsulation inside polyaniline nanotubules [15].

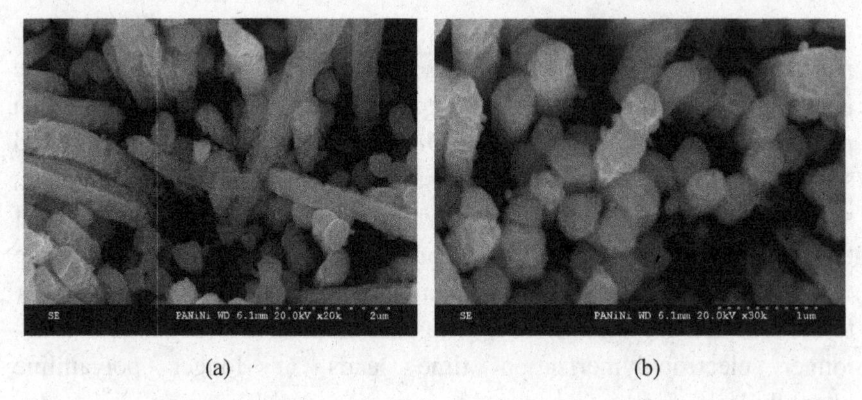

Figure 6.14 SEM micrographs of nickel nanowires encapsulated in polyaniline nanotubules [15]: (a) low magnification; (b) higher magnification.

To encapsulate nickel inside polyaniline nanotubules, silver film at the backside of AAO membrane is still the electrode for nickel electroplating process. Therefore nonconducting polyaniline nanotubules are required; otherwise, nickel will deposit on the surface of conducting polyaniline nanotubules. Chemical treatment was performed to convert conducting polyaniline nanotubules to nonconducting analogs. Nickel electroplating process was applied in the same single cell. After nickel electroplating and template etching, the released nanotubules were chemically treated by diluted hydrochloric acid. The hydrochloric acid treatment not only converts nonconducting polyaniline nanotubules back to conducting phase, also washes the possible deposited nickel particles on nanotubule surface for EDS element analysis. Figures 6.14 (a) and (b) show SEM micrographs of nickel nanowires encapsulated in polyaniline nanotubules. No hollow tubule was observed by SEM, compared with blank polyaniline nanotubules. Hence, it is assumed that something filled inside the nanotubules.

The existence of nickel can be confirmed by performing energy dispersive spectroscopy (EDS) on these nickel nanowires encapsulated in polyaniline nanotubules. Figure 6.15 is the EDS plot on nanotubules. Strong nickel peaks was observed in EDS spectrum, which indicates the presence of nickel element inside polyaniline nanotubules because any nickel particles on nanotubule surface have been already washed out by acid treatment after template etching.

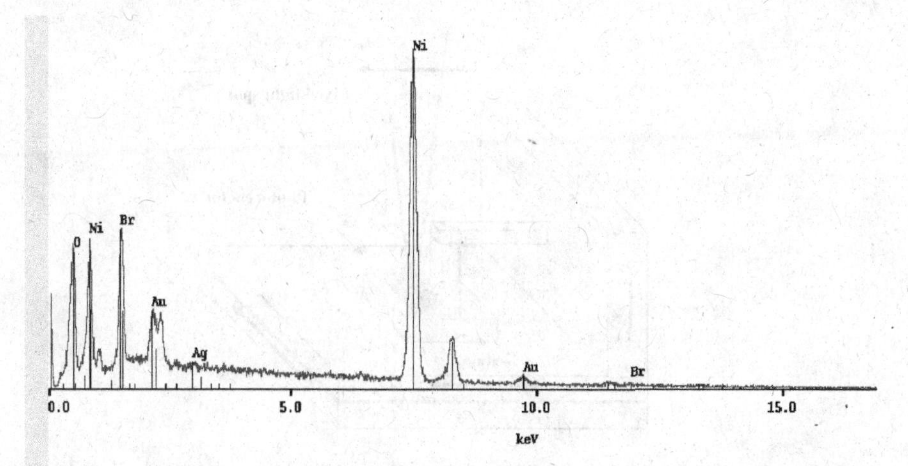

Figure 6.15 EDS spectrum of nickel nanowires encapsulated in polyaniline nanotubules [15].

6.4 Fabrication of 3D Polymer Microstructures

Three-dimensional polymer microstructures can be fabricated using microstereolithography (MSL) technique, which is derived from conventional stereolithography (SL). The microstereolithography shares the same principle with its macro scale counterpart, but in different dimensions. In MSL, a UV laser beam is focused to 1-2 μm to solidify a thin layer of 1-10 μm in thickness, while in conventional SL, laser beam spot size and layer thickness are both in the level of hundreds of microns. Submicron resolution of the x-y-z translation stages and the fine UV beam spot enable precise fabrication of real 3D complex microstructures using MSL.

6.4.1 *Microstereolithography*

MSL was used for the fabrication of high aspect ratio and complex 3D microstructure in 1993 [16, 17]. In contrast to conventional substractive micromachining, the MSL is an additive process, which enables one to fabricate high aspect ratio microstructures with novel smart materials. The MSL process is also, in principle, compatible with silicon process and batch fabrication is also feasible [18, 19].

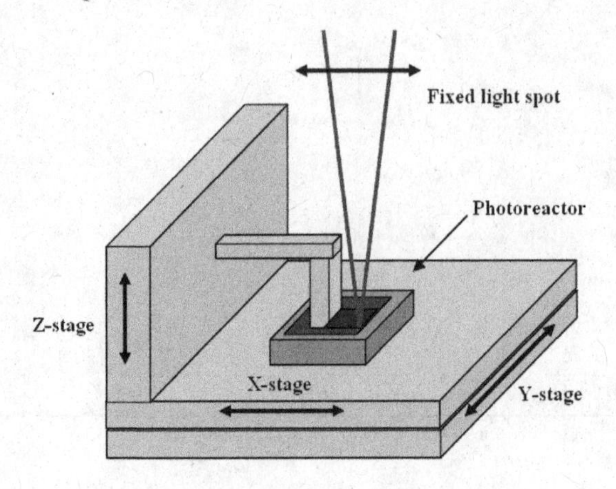

Figure 6.16 Microstereolithography with working piece scanning.

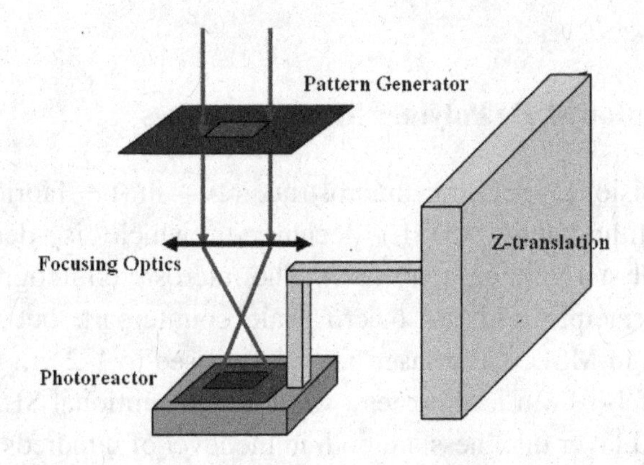

Figure 6.17 Microstereolithography with projection curing.

Different MSL systems have been developed aiming at precision and fabrication speed improvement. Basically, scanning MSL [16, 17, 20-22] and projection MSL [19, 23, 24] are the two major approaches. As shown in Figure 6.16, the scanning MSL builds solid microparts in a point-by-point and line-by-line fashion; while as shown in Figure 6.17 the projection MSL builds one layer with one exposure, saving time significantly.

Another effort in MSL development is to incorporate a broad spectrum of materials into fabrication for microelectromechanical systems (MEMS) with special functions. The MSL fabrication of micropolymeric parts and subsequently electro-plating of micrometallic parts have been explored [16-19, 21-24]. Functional polymer microparts possess the unique characteristics of high flexibility and low density, as well as electric conductivity in conductive polymers [16]. Microceramic structures have been fabricated by MSL from both structural ceramics and functional ceramics [20, 25]. Metal microparts fabricated by the similar photoforming process have also been realized [26]. Similar to the MSL of ceramic and metals, micro or even nano ceramic and metallic powders can be mixed with photocurable resin for the MSL of ceramics and metals.

6.4.2 *Two-photon microstereolithography*

As we know that conventional MSL has limitations in terms of the minimum thickness of the resin layers during the layer preparation due to viscosity and surface tension. The two photon MSL process, however, does not have this problem because the resin does not need to be layered. Usually, one photon is to be absorbed by a particle to produce photochemical change. But recently, a large number of experiments have been performed in which multiple photons are absorbed by a single particle for photochemical change, which means that single photon of energy is less than what is needed to cause a particle change unless in the case of high intensity. Multi-photon excitation is a nonlinear process, observed only with high intensities [27].

Two-photon absorption is one of the popular multiple photon excitations for photochemical change. There exist two different kinds of mechanisms for two-photon excitation. One is called sequential excitation, which involves a real intermediate state of the absorbing species. This intermediate state becomes very populated by the first photon, and it can act as the starting point for the absorption of the second one, as shown in Figure 6.18 (a). The real intermediate state A* has a well-defined lifetime, typically 10^{-4} to 10^{-9} s, so the second photon must be absorbed by the same particle within the lifetime of A* to cause the photochemical change. Because with sequential excitation the particle is excited by a resonantly absorbed photon, the overall sequential process is referred to as resonant two-photon excitation. The second

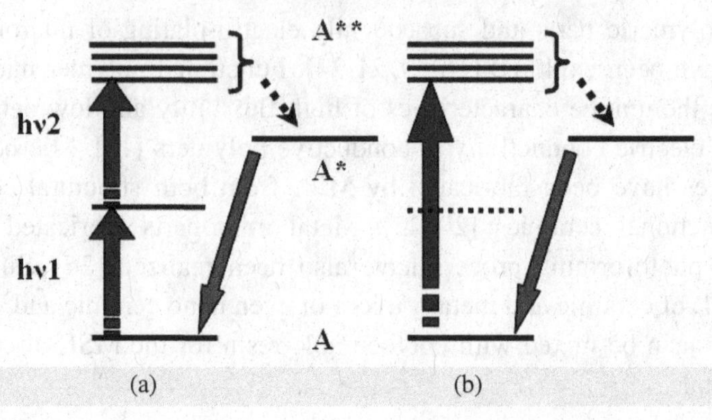

Figure 6.18 Two-photon excitation mechanisms: (a) resonant process and (b) non-resonant process.

excitation mechanism is nonresonant two-photon excitation, in which no intermediate states are involved, as shown in Figure 6.18 (b). A virtual state, A*, is created by the interaction of the first photon with particle A. Only if the second photon arrives during the duration of the first interaction (about 10^{-15}s) it can be absorbed. The process is also called simultaneous two-photon absorption. So it is apparent why high intensities are necessary for the two-photon excitation. In the resonant process, the second photon must be absorbed when the intermediate state A* is still alive. While in nonresonant process, two photons must reach the absorber within 10^{-15}s of each other [27]. Some microfabrications based on two-photon absorption have been reported [21, 28].

Specifically, when a laser beam is focused on a point with a microscope objective lens, as shown in Figure 6.19, photon density formed decreases with the distances away from focal plane, but the total photon number at every cross section remain constant, as shown in Figure 6.19 (b). The linear response of the materials to the light intensity based on single photon absorption does not have optical sectioning capability [29]. So, the resin is solidified completely in the illuminated region even beyond focal point, leading to a poor resolution. In contrast, if the material response is proportional to the square of the photon density, the integrated material response is enhanced greatly at the focal point because of the higher intensity, as shown in Figure 6.19 (c). So two-photon absorption based polymerization happens only in a small

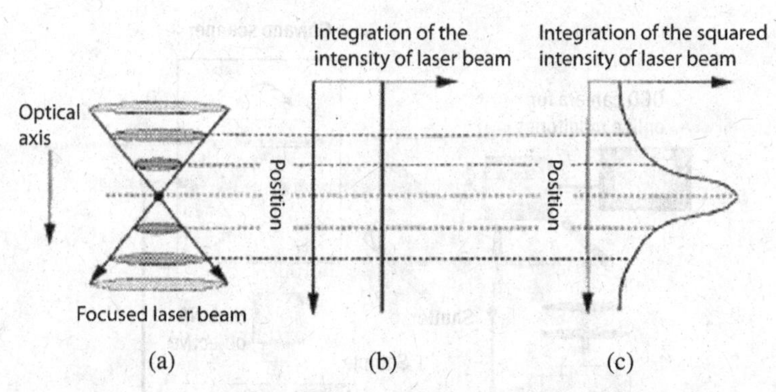

Figure 6.19 Comparison of two-photon absorption with one-photon absorption with the case of focused laser beam. (a) Focused laser beam, (b) total one-photon absorption per transversal plane, which is calculated by integrating the intensity over the plane, versus optical axis, and (c) total two-photon absorption per transversal plane, which is calculated by integrating the squared intensity over the plane, versus optical axis [29]. Source: Maruo, S. and Kawata S., (1998). Two-photon-absorbed near-infrared photopolymerization for three-dimensional microfabrication, *Journal of Microelectromechanical Systems*, 7, 411-415. © 1998 IEEE.

volume within the focal depth because it is associated with the square of the photon density. Normally, in order to obtain two photon absorption, the beam power of a laser has to be extremely high, usually several kilowatts.

Figure 6.20 shows a two-photon MSL apparatus. The beam from a mode locked titanium sapphire laser is directed to the galvanic scanning mirrors. It is then focused with an objective lens into the resin. A process monitor system including a camera is for the focus assistance and fabrication monitoring. A Z stage moves along the optical axis for multi-layer fabrication. From [10], the objective lens with a numerical aperture of 0.85 (magnification of 40) was used. The accuracy of the galvano-scanner set (general scanning) and the dc motor (sigma optics) scanner was 0.3 and 0.5 μm, respectively. The beam power at peak in the resin was about 3 kW with a repetition of 76 MHz and pulse width of 130 fs at a wavelength of 770 nm. The resin used was SCR-500, which is a mixture of urethane acrylate oligomers/monomers and photoinitiators, and the absorption spectrum of this resin is shown in Figure 6.21. The resin is transparent at 770 nm, meaning that no polymerization happens by one photon absorption.

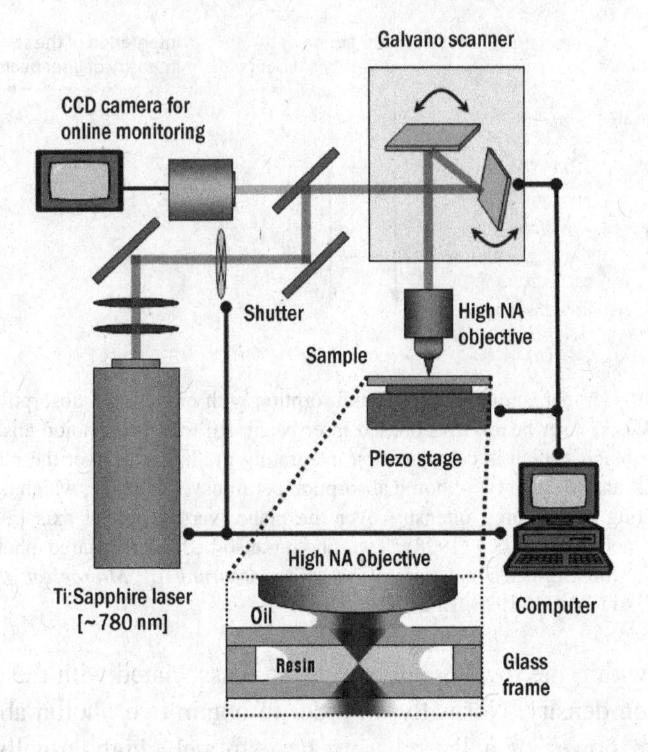

Figure 6.20 Schematic diagram of 3D microfabrication with two-photon absorption.

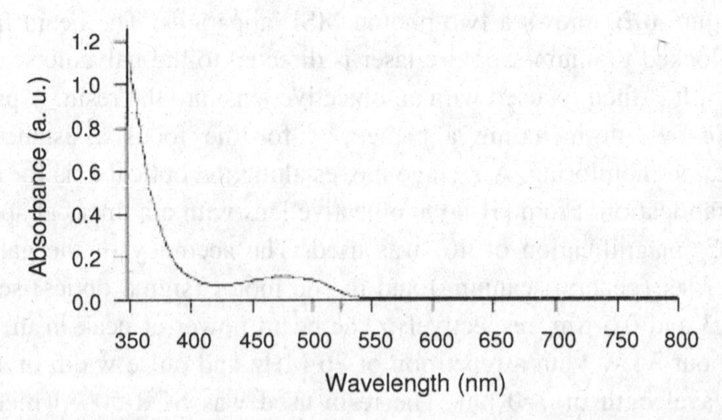

Figure 6.21 Absorption spectrum of the resin used for two-photon absorption MSL [29]. Source: Maruo, S. and Kawata S., (1998). Two-photon-absorbed near-infrared photopolymerization for three-dimensional microfabrication, *Journal of Microelectromechanical Systems*, **7**, 411-415. © 1998 IEEE.

Figure 6.22 shows some high aspect ratio and extremely fine 3D microstructures fabricated using two photon MSL [19]. The lateral and depth resolutions obtained with the two photon MSL are 0.62 and 2.2 μm, respectively. The depth resolution of 2.2 μm is hardly obtained with conventional MSL. The longest total length of a structure in the direction of the optical axis was up to 74.3 μm, which is a limitation of two photon MSL. In addition, the system is more expensive than most of the regular MSL systems.

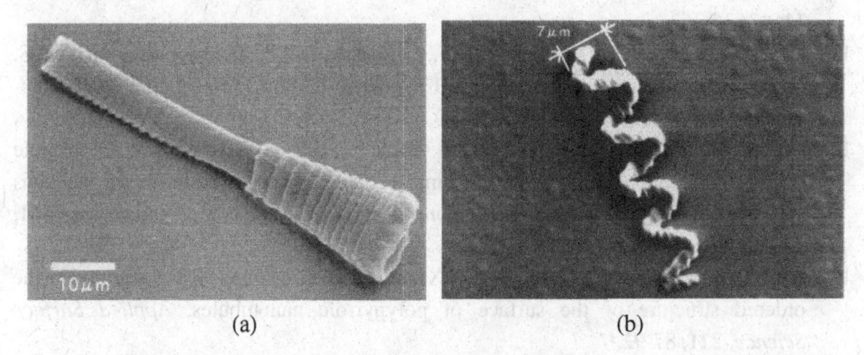

Figure 6.22 SEM images of 3D microstructures fabricated two-photon MSL: (a) microfunnel with length of 63 μm, and (b) microspiral structure [29]. Source: Maruo, S. and Kawata S., (1998). Two-photon-absorbed near-infrared photopolymerization for three-dimensional microfabrication, *Journal of Microelectromechanical Systems*, **7**, 411-415. © 1998 IEEE.

References:

1. Bar-Cohen, Y. (2004). *Electroactive Polymer (EAP) Actuators as Artificial Muscles: Reality, Potential, and Challenges*, SPIE Press.
2. Dubey, R., Shami, T. C., Rao, K. U. B., Yoon. H. and Varadan, V. K. (2009). Synthesis of polyamide microcapsules and effect of critical point drying on physical aspect, *Smart Materials and Structures*, **18** (2), 025021.
3. Soto-Portas, M. L., Argillier, J. F., Mechin, F. and Zydowicz, N. (2003). Preparation of oily core polyamide microcapsules via interfacial polycondensation, *Polymer International*, **52** (4), 522-527.
4. Yea, Y. and Park, K. (2004). Characterization of reservoir-type microcapsules made by the solvent exchange method, *AAPS PharmSciTech*, **5** (4), article no. 52.
5. Iijima, S. (1991). Helical microtubules of graphitic carbon, *Nature*, **354**, 56-58.
6. Hulteen, J. C. and Martin, C. R. (1997). A general template-based method for the preparation of nanomaterials, *Journal of Materials Chemistry*, **7**, 1075-1087.

7. Cai, Z., Lei, J., Liang, W., Menon, V. and Martin, C. R. (1991). Molecular and supermolecular origins of enhanced electronic conductivity in template-syntheiszed polyheterocyclic fibrils 1. Supermolecular effects, *Chemistry of Materials*, **3**, 960-967.
8. Martin, C. R. (1994). Nanomaterials – a membrane-based synthetic approach, *Science*, **266**, 1961-1966.
9. Martin, C. R. (1996). Membrane-based synthesis of nanomaterials, *Chemistry of Materials*, **8**, 1739-1746.
10. Cepak, V. M. and Martin, C. R. (1999). Preparation of polymeric micro- and nanostructures using a template-based deposition method, *Chemistry of Materials*, **11**, 1363-1367.
11. Delvaux, M., Duchet, J., Stavaux, P. Y., Legras, R. and Demoustier -Champagne, S. (2000). Chemical and electrochemical synthesis of polyaniline micro- and nano-tubules, *Synthetic Metals*, **113**, 275-280.
12. Qu, L., Shi, G., Yuan, J., Han, G. and Chen, F. (2004). Preparation of polypyrrole microstructures by direct electrochemical oxidation of pyrrole in an aqueous solution of camphorsulfonic acid, *Journal of Electroanalytical Chemistry*, **561**, 149-156.
13. He, J., Chen, W., Xu, N., Li, L., Li, X. and Xue, G. (2004). SERS studies on the ordered structure of the surface of polypyrrole nanotubules, *Applied Surface Science*, **221**, 87-92.
14. Cao, H., Xu, Z., Sheng, D., Hong, J., Sang, H. and Du, Y. (2001). An array of iron nanowires encapsulated in polyaniline nanotubules and its magnetic behavior, *Journal of Materials Chemistry*, **11**, 958-960.
15. Xie, J. and Varadan, V. K. (2005). Aligned blank and metal encapsulated conducting polymer nanotubules, *Proceedings of SPIE*, **5763**, 139-149.
16. Ikuta, K. and Hirowatari, K. (1993). Real three dimensional microfabrication using stereo lithography and metal molding, *Proceedings of IEEE MEMS' 93*, 42-47.
17. Katagi, T. and Nakajima, N. (1993). Photoforming applied to fine machining, *Proceedings of IEEE MEMS' 93*, 173-178.
18. Ikuta, K., Ogata, T., Tsubio, M. and Kojima, S. (1996). Development of mass productive micro stereo lithography (Mass-IH process), *Proceedings of IEEE MEMS' 96*, 301-305.
19. Bertsch, A., Zissi, S., Jezequel, J. Y., Corbel, S. and Andre, J. C. (1997). Microstereolithography using a liquid crystal display as dynamic mask-generator, *Microsystem Technologies*, 42-47.
20. Zhang, X., Jiang, X. N. and Sun, C. (1999). Micro-stereolithography of polymeric and ceramic microstructures, *Sensors and Actuators A*, **77**, 149-156.
21. Maruo, S. and Kawata, S. (1997). Two-photon-absorbed photopolymerization for three-dimensional microfabrication, *Proc. IEEE MEMS'97*, 169-174.

22. Zissi, S., Bertsch, A., Jezequel, J. Y., Corbel, S. Andre, J. C. and Lougnot, D. J. (1996). Stereolithography and microtechnologies, *Microsystem Technologies*, 97-102.
23. Nakamoto, T. and Yamaguchi, K. (1996). Consideration on the producing of high aspect ratio micro parts using UV sensitive photopolymer, *Proceedings of the Seventh International Symposium on Micro Machine and Human Science*, 53-58.
24. Monneret, S., Loubere, V. and Corbel, S. (1999). Microstereolithography using a dynamic mask generator and a non-coherent visible light source, *Proceedings of SPIE*, **3680**, 553-561.
25. Jiang, X., Sun, N. C. and Zhang, X. (1999). Micro-stereolithography of 3D complex ceramic microstructures and PZT thick films on Si substrate, *ASME MEMS*, **1**, 67-73.
26. Taylor, C., Cherkas, S. P., Hampton, H. J., Frantzen, J. B., Shah, O., Tiffany, W., Nanis, B. L., Booker, P., Salahieh, A. and Hansen, R. (1994). A spatial forming a three dimensional printing process, *Proceedings of IEEE MEMS'94*, 203-208.
27. Wayne, B. P. (1998). *Principles and Applications of Photochemistry*, Oxford University Press.
28. Cumpston, B. H., Ehrlich, J. E., Erskine, L. L., Heikal, A. A., Hu, Z. Y., Lee, L. Y. S., Levin, M. D., Marder, S. R., McCord, D. J., Perry, J. W., Rockel, H., Rumi, M. and Wu, X. L. (1998). New photopolymerization based two-photon absorbing chromospheres and application to three-dimensional microfabrication and optical storage, *Proceedings of Materials Research Society Symposium*, **488**, 217-225.
29. Maruo, S. and Kawata S., (1998). Two-photon-absorbed near-infrared photopolymerization for three-dimensional microfabrication," *Journal of Microelectromechanical Systems*, **7**, 411-415.
30. Kondo, H. (2008). What we have learned and will learn from cell ultrastructure in embedment-free section electron microscopy, *Microscopy Research and Technique*, **71**, 418-442.

Chapter 7

Nanocomposites

It has been a relentless endeavor of mankind to improve existing technologies and materials to attain higher perfection. Many exotic structural materials have been developed and among them one of the most promising materials is traditional advanced composites. This class of materials includes a wide variety of combination of reinforcements and matrices, and the usual reinforcements have the smallest dimension in micrometer. The most convincing example of such materials is naturally occurring structures, such as bones, which are natural composites having collagen fibers as reinforcements and hydroxyapatite as matrices.

With the advent of nanotechnologies, the interest of scientists and engineers has enormously increased in the development of nanocomposites which have one phase, usually reinforcing phase, on nanometric scale. Like traditional advanced composites, they have a variety of material options in reinforcement and matrices. Among them, the most common are metallic, polymeric, inorganic/ceramics materials. The concept of multiple-phase nanocomposites is not recent. This concept has been practiced ever since civilization started and humanity began producing more efficient materials for functional purposes. An excellent example of the use of synthetic nanocomposites in antiquity is the recent discovery of the constitution of Mayan paintings developed in the Mesoamericas. State-of-the-art characterization of these painting samples reveals that the structure of the paints consisted of a matrix of clay mixed with organic colorant (indigo) molecules. They also contained inclusions of metal nanoparticles encapsulated in an amorphous silicate substrate, with oxide nanoparticles on the substrate.

Nanocomposites have unprecedented combination of properties [1]. Among many properties, physical, mechanical, chemical, thermal and optical properties are of much interest. Nanoreinforced films are transparent and tough, and they have brought a big change in the packaging industry. The properties of a nanocomposite largely depend on its reinforcement and matrix. By using heterogeneous chemical species as reinforcements and matrices, nanocomposites with different properties can be obtained, and such nanocomposites may be multi-functional. Similar to traditional composites, the interphase between the reinforcement and the matrice is very important for a nanocomposite. Moreover, due to their nanometer size, proper functionalization and compatibility of reinforcements are often required; otherwise, the nanoscale reinforcement phase may easily agglomerate or become segregated from the bulk matrix, resulting in poor properties.

Due to their extensive engineering applications, the research on nanocomposites has obtained strong supports from governments and industries all over the world, and has shown great market expectations. The global market for nanocomposites is about $250 million by 2008, with annual growth rates projected to be 18-25% per year.

7.1 Nanoscale Reinforcements

In a nanocomposite, nanoscale reinforcements usually present 1-5% by weight of the matrix. Due to high surface area of nanoscale materials, this small quantity of nanoscale reinforcement is sufficient to drive maximum properties. Higher loading may have adverse effects on nanocomposites due to phase segregation and agglomeration. Nanoscale reinforcements are broadly divided into three categories: nanoscale carbons, nanoclays and equi-axed nanoparticles.

7.1.1 *Nanoscale carbons*

The most prominent reinforcing material in this category is carbon nanotubes (CNTs). Other materials used include fullerenes, carbon nanospheres and graphite powders [2-12]. A detailed description of CNTs and fullerenes has been given in Chapter 4 of this book.

As good interphase is important for composites and carbon is chemically inert to interact with matrix, the functionalization of nanoscale carbon is needed. Carbon nanofibers have yielded promising results in nanocomposites for structural applications.

7.1.2 *Nano clays*

Nano clays are a kind of inorganic nanomaterials which are very promising due to their low cost and moderate thermal stability. Nano clays are minerals which have layered structures. The most important member of the nano clay family is montmorillonite. This is a major constituents of bentonite which belongs to smectite family. Montmorillonite consists of two tetrahedral silica (corner shared) sheets attached to one octahedral (edge shared) alumina or magnesia sheet. The thickness of these layers is about 1 nm and other dimensions of these layers may vary from a few nanometers to a few micrometers. This is the reason why nano clay is called two-dimensional (2d) material. The structure of montmorillonite is depicted in Figure 7.1.

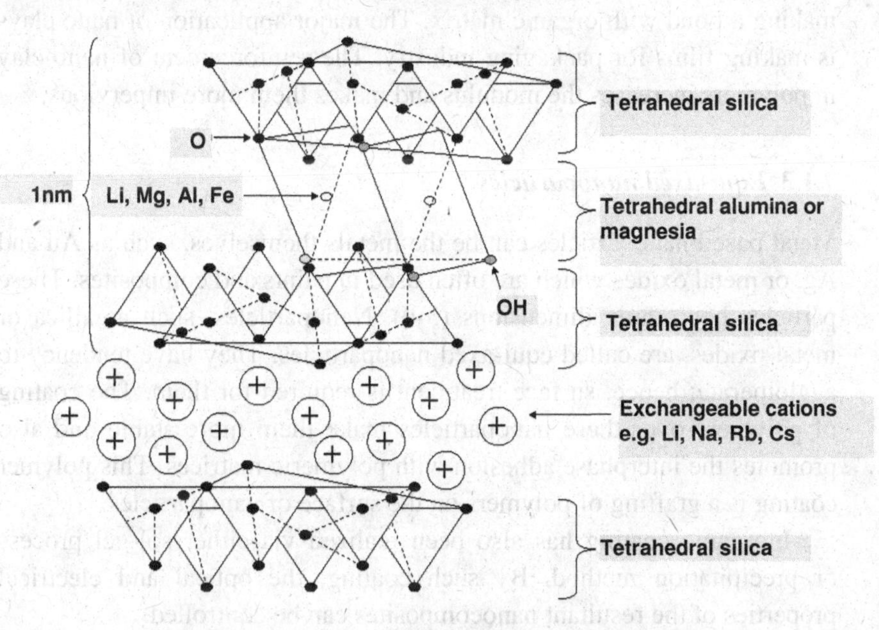

Figure 7.1 Structure of nano clay montmorillonite.

Each unit cell of a montmorillonite nano clay has a negative charge between 0.6 to 1.3 eV due to isomorphic substitution of alumina (Al^{3+}) into silicate (Si^{4+}) or interaction of magnesia (Mg^{2+}) and alumina (Al^{3+}). These layers are held together by the charge compensation cations, such as Li^+, Na^+, Rb^+ and Cs^+.

Montmorillonite nano clays are often treated with talo, animal fat. The application of talo on nano clays is only for handling and storage purposes. The treatment with talo causes an attachment of olefinic chain on the surface of nano clays so that they do not agglomerate. The montmorillonite nano clays are hydrophilic, while for the good interphase adhesion within polymeric matrices, hydrophobicity is required. The initial treatment of talo is not sufficient to provide enough hydrophobicity for interphase interaction.

Therefore, nano clays are usually treated with n-dodecylamine ammonium salt, 12-aminododecanoic acid ammonium salt, dimethyl-octadyl ammonium chloride (DODAC), cetyltrimethylammonium bromide (CTAB). In the treatment with these salts, the ammonium group is attached to nano clays leaving the long olefinic chain suspended for making a bond with organic matrix. The major application of nano clays is making films for packaging industry. The reinforcement of nano clay in polymers increases the modulus and makes them more impervious.

7.1.3 *Equi-axed nanoparticles*

Metal based nanoparticles can be the metals themselves, such as Au and Ag, or metal oxides which are often used in paints and composites. These particles have three dimensions (3-d). Nanoparticles, such as silica or metal oxides, are called equi-axed nanoparticles. They have tendency to agglomerate; hence, surface treatment is required for them. The coating of polymers over these nanoparticles make them more stable and also promotes the interphase adhesion with polymeric matrices. This polymer coating is a grafting of polymers on the surface of nanoparticles.

Inorganic coating has also been realized via either sol-gel process or precipitation method. By such coating, the optical and electrical properties of the resultant nanocomposites can be controlled.

7.2 Ceramic Matrix Nanocomposites

Conventional ceramics suffer from low damage tolerance, low creep and fatigue, low thermal shock resistance and so on. Therefore, efforts have been made to solve these problems. The reinforcement in ceramics could overcome these problems to some extent [13, 14]. Moreover, with nanoscale reinforcement, the microstructure of ceramics is maintained. Hence, its advantage is that ceramics do not lose their surface finish and luster. These composites are based on passive control of the micro-structures by incorporating nanometer-size second-phase dispersions into ceramic matrices.

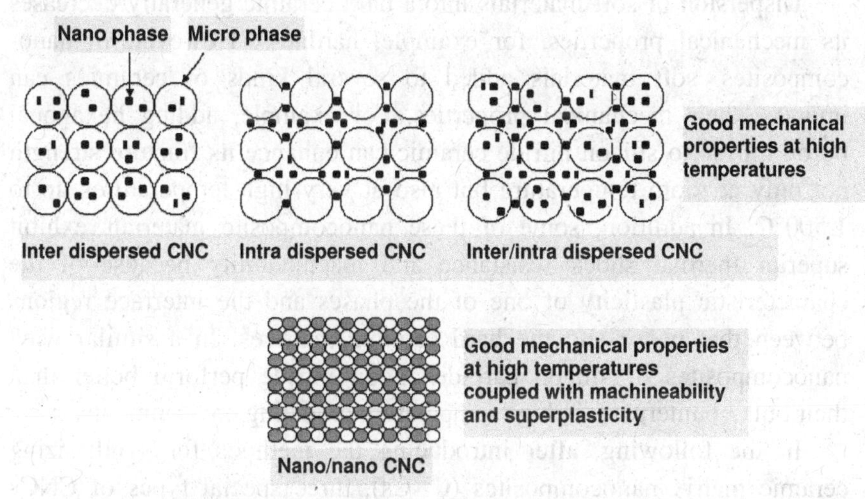

Figure 7.2 Effect of nanoscale reinforcements in CNCs.

These materials can be produced by incorporating a very small amount of additive into a ceramic matrix. The reinforcement segregates at the grain boundary with a gradient concentration, or precipitates as molecular or cluster sized particles within the grains or at the grain boundaries. The dispersion of nanophase in ceramics is termed as intergranular, intragranular or the combination of the both. This has been shown schematically in Figure 7.2. Optimized processing can lead to excellent structural control at the molecular level in most nanocomposite materials. Intergranular dispersions control the shape and size of the grains of ceramic matrix, whereas intragranular dispersions control the grain boundry of the ceramic matrix. This results in improvement in mechanical properties. Nanocomposite technology is also applicable to functional ceramics such as ferroelectric, piezoelectric, varistor and ion-conducting materials. Incorporating a small amount of ceramic or metallic nanoparticles into Al_2O_3, $BaTiO_3$, ZnO or cubic ZrO_2 can significantly improve their mechanical strength, hardness and toughness, which are very important in creating highly reliable electric devices operating in severe environmental conditions. In addition, dispersing conducting metallic nanoparticles or nanowires can enhance the electrical properties.

Dispersion of soft materials into a hard ceramic generally decreases its mechanical properties, for example, hardness. However, in nano-composites, soft materials added to several kinds of ceramics can improve their mechanical properties. For example, adding hexagonal boron nitride to silicon nitride ceramic can enhance its fracture strength not only at room temperature but also at very high temperatures up to 1,500°C. In addition, some of these nanocomposite materials exhibit superior thermal shock resistance and machinability because of the characteristic plasticity of one of the phases and the interface regions between that phase and the hard ceramic matrices. In a similar way, nanocomposites of silicon carbide/silicon nitride perform better than their bulk counterparts in high temperature oxidizing environments.

In the following, after introducing the methods for synthesizing ceramic matrix nanocomposites (CNCs), three special types of CNCs will be discussed: CNT reinforced CNCs, carbon-carbon nanocomposites and thin film nanocomposites.

7.2.1 *Synthesis methods*

Many methods can be used for the synthesis of CNCs. In the following, we discuss sol-gel method, chemical vapor deposition method, mechanical applying method and thermal spraying method. More information about the first two methods, i.e. sol-gel method and chemical vapor deposition method, can be found in Chapter 3 of this book.

7.2.1.1 *Sol-gel processing*

CNS based on heterometallic ceramics may be processed by this method. This method is useful for films and epitaxially oriented materials which have specific applications in electronics and optoelectronics.

7.2.1.2 *Chemical vapor deposition*

This is the most popular method for preparing CNCs. Through this method, the chemical composition and structure can be controlled at atomic and molecular levels due to the decomposition of gases at high temperature. In this method, both the reinforcement and matrix phases are co-deposited on a substrate to give CNC. The structures of CNCs can be controlled by controlling the reaction conditions, so this method is useful for synthesizing functionally graded CNCs. A variety of compositions of nitrides, carbides silicides and oxides have been synthesized by this method. These nanoreinforcements interact with the ceramic matrix, by inter or intra dispersion, and cause restrictions to dislocation movements. This leads to enhanced mechanical strength of CNCs.

7.2.1.3 *Mechanical alloying*

The ball milling process has been one of the conventional techniques of making CNCs. In this process, ball milling is used to mix and homogenize the constituents for making CNCs. As shown in Figure 7.3,

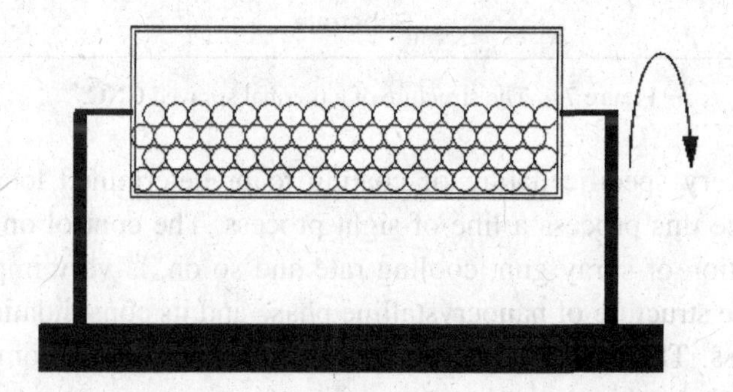

Figure 7.3 Schematic diagram of a ball milling system.

a ball milling system contains steel balls in a container rotating at a high speed. The balls inside the container hammer and crush the constituents, inducing chemical reactions such as displacement reactions. For example, reaction of aluminum with some metal oxides results in alumina.

7.2.1.4 *Thermal spraying*

Thermal spraying is a very popular commercial method for the production of hard coatings of CNCs. In this process, the reinforcement material is melted and sprayed on a substrate, matrix material. The molten material sprayed at a high speed flattens and cools down at moment it comes in contact with the substrate. The bond between the sprayed particles and substrate is mechanical locking, but it can be improved by the control of the speed of impinging particles, the cleanliness of the substrate surface and the diffusion treatment to the impinging particles. As schematically shown in Figure 7.4, in a CNC synthesized by thermal spraying method, voids and oxides may present.

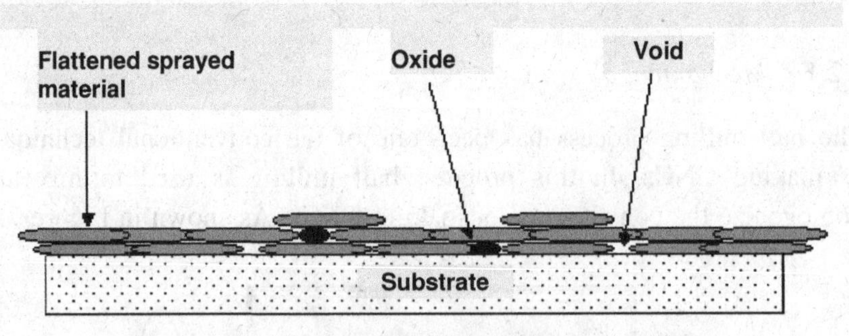

Figure 7.4 The structure of a thermal sprayed CNC.

The very specific nature of coating to predetermined location and areas make this process a line-of-sight process. The control on gas flow rate, location of spray gun, cooling rate and so on, is very important to control the structure of nanocrystalline phase and its consolidation during the process. The hard coating of oxides, carbides or their combinations can be made by this method.

There are mainly three types of thermal spraying methods: flame spraying method, plasma spraying method and high speed oxyfuel method. In a flame spraying method, oxyacetylene flame is used for melting and impingement. In a plasma spraying method, ionized gas plasma is used as a heating and impingement source. In a high speed oxyfuel method, a mixture of hydrogen and oxygen is the medium of heating and spraying. There some other methods, such as low pressure plasma spraying, vacuum plasma spraying and air plasma spraying, are variants of above mentioned methods.

7.2.2 *CNT reinforced CNCs*

Due to the attractive properties of carbon nanotubes (CNTs), CNT reinforced CNCs exhibit advantages over the conventional composite materials. They are extensively investigated and can be widely used for various engineering purposes. CNT reinforced CNCs are especially important for structural applications, and they have great application potentials in the aerospace industry.

The attractive properties of CNT reinforced CNCs are based on the assumption that the CNT reinforcements are uniformly distributed in the ceramic matrices. Therefore, in the fabrication of CNT reinforced CNCs, special attention should be paid on how to ensure the uniform dispersion of CNTs in the ceramic matrices, which is the key to good CNTs reinforced CNCs.

The sintering method is often used in the fabrication of CNT reinforced CNCs. In this method, carbon nanotubes are mechanically mixed with ceramic materials and then hot pressed. Two methods are often used in mixing CNTs with ceramic matrices. One is the sol-gel method, in which CNTs are dispersed in a sol, and the other is the CVD method in which CNTs are grown on a ceramic substrate.

The reinforcement of CNTs in ceramics enhances the physical and mechanical properties of the ceramics. Because of their high aspect-ratio and other mechanical properties, CNTs are flexible, so the reinforcement of CNTs greatly improves the fracture toughness of the resultant nanocomposites. An increase in fracture toughness on the order of 10% has been obtained in carbon nanotube/nanocrystalline SiC ceramic composites which were fabricated by a hot-pressing method at 2273 K (25 MPa in Ar for 1 h). Meanwhile, due to their ceramic matrices, CNT reinforced CNCs usually have inherently smooth surfaces, so they are self-lubricating.

The reinforcement of CNTs changes the electrical properties of the ceramic matrices used. If the amount of reinforced CNTs is higher than a critical value, the good electrical conductivity of CNTs converts the insulating ceramics to conducting nanocomposites. Besides, due to the quasi-one dimensional structure of CNTs, when CNTs reinforced CNCs are extruded, the CNTs reinforced in the composite get aligned in the direction of flow, and this causes anisotropy to the isotropic ceramics.

7.2.3 *Thin film nanocomposites*

Thin film nanocomposites are films consisting of more than one phases, and at least one of the phases have a dimension in the nanometric range. In the following, we discuss the synthesis, properties and applications of two typical types of thin film nanocomposites: granular films and multi-layer films.

7.2.3.1 *Granular films*

Granular nanocomposite films are those that contain both phases (metal and ceramic) in the same layer of the film and have no abrupt interfaces across the film thickness. In granular nanocomposite coatings, usually the matrix phase is a polymer and the dispersed phase is inorganic. Granular films in which at least one distributed phase has electrical/magnetic properties are mainly used in electrical and magnetic applications. In certain systems, consisting of metal and ceramic particles (such as iron oxide/silver and alumina/nickel), changing the fractions of the phases present can alter their electrical and magnetic properties. At small volume fractions of the metal component, the material exhibits ferromagnetic behavior. Beyond a certain volume fraction of the metal phase, the ferromagnetic ordering gives way to superparamagnetism. An important parameter that critically affects the properties of granular films is the percolation threshold of the metal, which typically corresponds to a metal volume fraction of 0.5-0.6. Once the metal particles attain percolation at higher volume fractions, the film behaves essentially like the metallic phase.

Many methods, such as CVD and electrochemical methods, can be used to prepare homogeneous nanostructured composite films. For example, nanostructured AlN/TiN composite films can be made by CVD using high-speed deposition of gaseous precursors ($AlCl_3$, $TiCl_4$, NH_3) to form insoluble solid mixtures. The generated films have grain sizes 8 nm (AlN) and 6 nm (TiN). Compared to bulk AlN/TiN composites, these composite films have better ductility and greater toughness.

7.2.3.2 *Multi-layer films*

Multi-layered thin-film nanocomposites consist of alternating layers of different phases, and they have a characteristic thickness on the order of nanometers. These films are usually used for their enhanced hardness, elastic moduli and wear properties. The elastic moduli of multi-layered thin films are usually higher than those of homogeneous thin films of the components. The supermodulus effect is observed in some metallic systems, by which, at certain characteristic thicknesses of the film, the elastic modulus increases by more than 200%. The most satisfactory explanation of this effect is the incoherent interface between the adjacent layers. This interference causes the displacement of the atoms from their equilibrium positions and, during the loading, all the layers undergo compression, which results in a higher resistance to deformation. The hardness of these multi-layer films is quite good due to the stacking of ultrathin films and the dislocations present in them. These dislocations cannot move from one layer to another because of their different dislocation line lengths. Conventional thin-film deposition techniques, such as sputtering, physical vapor deposition, CVD and electrochemical deposition, can be used to produce multi-layer thin-film nanocomposites.

7.3 Metal Matrix Nanocomposites

Metal matrix nanocomposites (MMNCs) are a kind of composites in which rigid ceramic reinforcements are embedded in a ductile metal or alloy matrix. MMNCs have high specific strengths and very useful for special applications such as aerospace and adventure sports. For instance, magnesium has low density and moderate strength, but, after the reinforcement of nanoceramics, such as SiC and SiN, its strength for the same density improves significantly. It is the same for SiC reinforced aluminum. The most common way for producing metal matrix nanocomposites is sintering. The dispersion of nanophase is important to obtain optimum properties. Other important methods for producing MMNCs include ball milling and sputtering [15].

MMNCs have a wide range of matrix materials including aluminum, titanium, copper, nickel and iron, and the reinforcements can be borides, carbides, nitrides, oxides and their mixtures. Because of the formation of stable nano-sized ceramic reinforcements, MMNCs exhibit excellent mechanical properties. MMNCs combine metallic properties (ductility and toughness) with ceramic characteristics (high strength and modulus), leading to high specific modulus, high strength and high thermal stability.

Mainly due to following three reasons, MMNCs have been extensively used in the aerospace and automotive industries and other structural applications. First, various types of reinforcements with competitive costs are available. Second, due to the successful development of manufacturing processes, MMNCs can be produced with reproducible structure and properties. Third, the standard or near standard metal working methods are available for fabricating these composites.

The family of discontinuous nanomaterials reinforced in MMNCs includes both particulates and whiskers or short fibers. The particulate-reinforced MMNCs are of particular interest due to their ease of fabrication, lower costs and isotropic properties. Traditionally, discontinuously reinforced MMNCs have been produced by several processing routes such as powder metallurgy, spray deposition, mechanical alloying (MA) and various casting techniques. All these techniques are based on the addition of ceramic reinforcements to the matrix materials, which may be in molten or powder form.

7.4 Magnetic Nanocomposites

Magnetic nanocomposites have revolutionized the electronics and optoelectronics industries. These nanocomposites have given a new dimension to data storage devices. In the following, we discuss particle-dispersed magnetic nanocomposites and magnetic multi-layer films.

7.4.1 *Particle-dispersed magnetic nanocomposites*

In a particle-dispersed magnetic nanocomposite, the magnetic species are dispersed within nonmagnetic or magnetic matrices. Compared to conventional magnetic materials, magnet nanocomposites have higher remanence and large energy product. Such nanocomposites have many practical applications.

Different synthesis routes can be used to produce magnetic composites. Alloy ribbons of nanocrystalline and nanocomposite magnets, for instance, containing $Nd_2Fe_{14}B$, Fe_3B and α-Fe, can be prepared by melt spinning/splat cooling of alloys accompanied by heat treatments such as fast annealing. The most successful technique has been mechanical alloying of two-phase mixtures, such as α-$Fe/Sm_2Fe_{17}N_3$, which typically results in two-phase mixtures consisting of a nanocrystalline soft magnetic matrix and an amorphous hard magnetic phase. Annealing and crystallization at higher temperatures lead to nanocomposite magnets that show superior magnetic behaviors. Powder synthesis process is another process for the fabrication of magnetic nanocomposites. In this technology magnetic nanoparticles are coated with insulating materials and these coated particles are consolidated into exchange coupled cores.

7.4.2 *Magnetic multi-layer films*

Nowadays, the storage density of hard disk drives increases quickly, and meanwhile the cost per megabyte of storage decline rapidly. Along with disk performance and price, the sputtering equipment is evolving in both design and efficiency. Standard processing modules involve heating, DC or RF sputtering deposition and cooling.

The electrical resistance of a magnetic multi-layer film may change greatly with the application of an external magnetic field, and this effect is usually called giant magnetoresistance (GMR). Various types of magnetic sensors have been developed based on GMR effect, and applied in biomedical engineering [16].

7.5 Polymeric Nanocomposites

Polymeric nanocomposites (PNCs) have overcome the limitations of traditional micrometer-scale polymeric composites and opened up a large window of opportunities [17-24]. In polymeric nanocomposites, the filler is less than 100 nm in at least one dimension. Mainly due to following three reasons, research and development of nanofilled polymers have been very active. First, unprecedented combinations of properties have been observed in some polymer nanocomposites. Expected benefits from nanocomposites include improvement in modulus, flexural strength, heat distortion temperature and barrier properties. Unlike typical mineral reinforced systems, they do not have the conventional trade-offs in impact and clarity. Nanocomposites can be optically transparent and optically active. The second reason is the discovery and production of carbon nanotubes, and their applications in polymeric nanocomposites. Third, due to the significant development in the chemical processing of nanoparticles and *in situ* processing of nanocomposites, the morphology of such composites can be well controlled, and the interface between the matrix and the reinforcements can be controlled in an unprecedented way.

The polymeric nanocomposites are different from conventional composites [25]. The comparison has been tabulated in Table 7.1. As shown in Figure 7.5, the comparison tabulated may be testified and correlated with the electron micrographs of conventional composites and polymeric nanocomposites [26].

One of the key limitations in the commercialization of nanocomposites is the processing techniques. The most challenging task in the development of PNCs is the proper dispersion of the nanoreinforcement in the matrix. Without proper dispersion and distribution of the reinforcement, the high surface area is compromised and the aggregates can act as defects, which limit properties. Distribution of nano-reinforcement describes the homogeneity throughout the sample, and the dispersion describes the level of agglomeration. To achieve good performances, good distribution and good dispersion are required.

Table 7.1 Comparison between conventional polymeric composites and polymeric nanocomposites.

Conventional Polymeric Composites	**Polymeric Nanocomposites**
Conventional laws of rule of mixture are readily applied	Conventional laws of rule of mixture are not applied
Moderate interfacial areas	Large interfacial areas
Bulk matrix exists around the reinforcement	Size of the interphase region comparable to the size of the reinforcement and matrix
Interconnected network of interfacial polymer	Restrictions on chain conformations

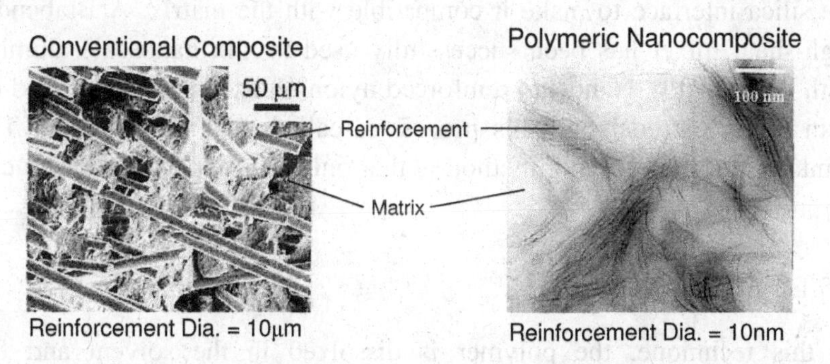

Figure 7.5 Comparison between conventional polymeric and polymeric nanocomposites. Source: Gacitua, W. E., Ballerini, A. A. and Zhang, J. (2005). Polymer nanocomposites: synthetic and natural fillers – a review, Maderas Ciencia y Tecnología, 7(3), 159-178.

In the following, after introducing the typical methods for synthesizing PNCs and the characteristic properties of PNCs, four types will be discussed: clay reinforced PNCs, CNT reinforced PNCs, rubber matrix nanocomposites and stealth PNCs.

7.5.1 *Synthesis of polymeric nanocomposites*

7.5.1.1 *Melt mixing method*

In this method, nanoreinforcements and thermoplastic polymer matrix are mixed in two roll mill, twin screw extruder or Brabender high shear mixer. In all these techniques, different types of shearing elements are used to achieve homogeneous mix in which nanoparticles are dispersed to the maximum extent. In the mixing process, the polymer is heated so that viscosity comes down and shearing elements disperse the nano-reinforcement. This pronged approach of melt mixing process causes efficient dispersion of nanoreinforcement. For example, polypropylene and nanoscale silica have been mixed successfully in a two-roll mill, but samples with more than 20 wt.% filler could not be drawn. This is typical and is a limitation of this kind of processing method. Nanoscale silica/polypropylene composites have been processed in a twin-screw extruder, but the dispersion was successful only after the modification of the silica interface to make it compatible with the matrix. A Brabender high-shear mixer has been successfully used to mix nanoscale alumina with PET, LDPE. Nanoclay reinforced nylon-6 can also be processed by twin screw extruder and this process is called melt intercalation. The limitation of melt mixing method is that only thermoplastic rubbers can be processed.

7.5.1.2 *Solution mixing*

In this technique, the polymer is dissolved in the solvent and the nanoparticles are dispersed in the solution. The surface of nanoparticles is modified by the treatment with suitable chemical, making them compatible with polymeric solution. The nanoparticles are properly mixed, preferable by ultrasonication, in order to make nanoparticles disperse completely. The nanoparticle/polymer solution can then be cast into a solid by solvent evaporation or precipitation. Further processing can be done by conventional techniques.

7.5.1.3 *In-situ polymerization*

In this technique, nanoscale particles are dispersed in a monomer or monomer solution, and the resulting mixture is polymerized by standard polymerization methods [27]. One fortunate aspect of this method is the potential to graft the polymer onto the particle surface. Many different types of nanocomposites have been processed by *in-situ* polymerization. A few examples are silica/ Nylon6, silica/ poly 2-hydroxyethylmethacrylate, alumina/ polymethylmethacrylate, titania/ PMMA and $CaCO_3$/ PMMA. The key to *in-situ* polymerization is the appropriate dispersion of the reinforcement in the monomer. This often requires modification of the particle surface so that dispersion is easier.

7.5.1.4 *In-situ particle processing*

An interesting method for producing nanoparticle-filled polymers is *in-situ* sol–gel processing of the particles inside a polymer. The process has been successfully used to produce polymer nanocomposites with silica and titania in a range of matrices. The overall reaction for silica from tetrethylorthosilicate (TEOS) is shown below.

$$Si(OC_2H_5)_4 + \text{excess } H_2O = SiO_2 + 4C_2H_5OH \qquad (7.1)$$

Metal/polymer nanocomposites can also be processed via *in-situ* formation of metal particles from suitable metal precursors. The reaction occurs in the presence of a protective polymer, which limits the size of the particles. Once a stable suspension of metal particles is prepared in the presence of a polymer, the composite can be casted.

7.5.1.5 *Thermal spray method*

This method is explained in detail in Section 7.2 of this chapter. The only difference over here is, in place of ceramics, polymers are used. Thermal spraying has also been successful in processing nanoparticle reinforced Nylon.

7.5.1.6 *Template method*

This is a solution based processing of PNCs. In this technique, either monomer is inserted inside the nanoscale pores or layers of inorganic host material and polymerized, or nucleation of inorganic nanophase takes place in polymer solution. *In-situ* polymerization of pyrrole or thiophene in zeolite is the example of synthesis of PNCs through *in-situ* polymerization of monomeric precursor inside nanopores. The PNCs synthesized by this technique find applications in electronics/optoelectronics devices. For the hosts like aluminosilicates which have larger diameter channel ranging may accommodate more polymer per channel. This enables the growth of polymer at walls of mesopores and this leads to tubular polymer structures. This type of PNCs can be used in drug delivery applications.

7.5.2 *Typical properties of polymeric nanocomposites*

Polymers and polymeric composites are replacing the conventional materials at a very fast pace due to their cost effectiveness and tailorability to suit most of the applications in the industry. In the following, we discuss the typical properties of polymeric nanocomposites [1].

7.5.2.1 *Mechanical properties*

One of the primary purposes of reinforcing polymers is to improve their mechanical performances [28]. In traditional composites, unfortunately, this often comes at the cost of a substantial reduction in ductility, and sometimes, in impact strength. This owes to the stress concentrations caused by the fillers. Well-dispersed nanofillers, on the other hand, can improve the modulus and strength, and maintain or even improve ductility because their small size does not create large stress concentrations.

7.5.2.2 *Glass transition and relaxation behavior*

The interfacial region is extremely large in nanocomposites. The interaction of the polymer with the nanoparticles gives significant opportunity for changing the polymer mobility and relaxation dynamics. For example, polystyrene chains intercalated between the layers of a smectic nano clay have more mobility than those in a bulk polymer. This greater mobility may be due to an ordering that occurs between the layers, which create low- and high-density regions, thus providing the opportunity for mobility in the low-density regions. Though the specific mechanics of chain dynamics are not yet completely understood, it is very clear that the rheology/glass transition temperature of a polymer can be controlled by changing the polymer mobility with nanocomposite interfaces.

7.5.2.3 *Abrasion and wear resistance*

Although it is well known that the abrasion resistance of filled polymers depends on particle size, the incorporation of nanoscale fillers has led to unexpected results. Most of the reinforcing particles are stable and increase the abrasion resistance of the composite if reinforcing particles are larger than the abrasive particles. As the size of reinforcing particles is decreased to a size similar to that of the abrading particles, filler particles are removed, and the abrasion resistance is compromised. This does not happen with the nanoscale reinforcement. For example, the addition of nano $CaCO_3$ to PMMA results in a significant decrease (factor of 2) in material loss due to abrasion. In addition, nanoparticles can simultaneously improve wear resistance and decrease the coefficient of friction.

7.5.2.4 *Permeability*

The reduced gas and liquid permeability of nanoreinforced polymers makes them attractive membrane materials. The nano clay (plate like - 2d material) improves the permeability remarkably. The large change in permeability of liquid or gas through a composite material can be explained from simple predictions. For a plate-like material with an aspect ratio of L/W dispersed parallel in a matrix, its tortuosity factor S is given by

$$S = 1 + (L/2W)V_f \qquad (7.2)$$

where V_f is the volume fraction of plate-like filler. The relative permeability coefficient P_c/P_p is given by:

$$P_c/P_p = 1 / (1 + (L/2W)V_f) \qquad (7.3)$$

where P_c and P_p are the permeability coefficients of the composite and polymer matrix, respectively. A slightly more rigorous analysis for a random in-plane arrangement of plates yields

$$P_c/P_p = 1/[1 + I(L/W)^2(V_f^2/(1 - V_f))] \quad (7.4)$$

where I depends on the distribution of the plate-like material. In either case, the higher the aspect ratio of the filler, the larger the decrease in permeability.

7.5.2.5 *Dimensional stability*

Dimensional stability is critical in many applications. For example, if the layers of a microelectronic chip have different thermal or environmental dimensional stabilities, residual stresses can develop that may cause premature failure. Poor dimensional stability can also cause warping or other changes in shape that affect the function of a material. Nanocomposites improve both thermal and environmental dimensional stability.

7.5.2.6 *Thermal stability and flammability*

PNCs have good thermal stability as they have improved glass transition temperature. Due to its high surface area, nanoreinforcement binds the matrix at very small level, and hence the free translational movement is restricted. This causes the improvement in thermal stability. The increase of flammability resistance is due to the improved barrier properties of the composites. If oxygen cannot penetrate, it cannot cause oxidation of the resin. In addition, the inorganic phase can act as a radical sink to prevent polymer chains from decomposing.

The flammability resistance of clay-filled polymers indicates that their ablation resistance might also be excellent. As a material is heated during ablation, the surface of the material reacts and forms a tough char. If the char is not reinforced, it fails and is removed from the surface, exposing more material. Traditional composites require a significant weight fraction of filler (more than 30 wt.%) to achieve significant ablation resistance. On the other hand, when a piece of Nylon 6 filled with 2-5 wt.% nano clay is exposed to a mock solid rocket motor firing rig, a layer of char will be formed on the surface, significantly retarding further erosion. In addition, oxygen plasma forms a passivation layer on Nylon 6/ nano clay nanocomposites, which significantly retards further erosion of the composite surface. This behavior is not a strong function of the organic molecules used to modify the clay or the strength of the clay/polymer interaction, but is a function of the degree of exfoliation.

7.5.2.7 *Resistivity, permittivity and breakdown strength*

When the reinforcement of a polymeric composite gets to the nanoscale, the electrical properties of the composite are expected to be different mainly due to two reasons. First, quantum effects become important, so the electrical properties of nanoparticles can change compared to the bulk. Second, for the same volume fraction, the interparticle spacing decreases along with the decrease of particle size. Therefore, percolation occurs at lower volume fractions.

The permittivity of polymers can be increased with the addition of many metal oxide reinforcement at the nanoscale. Furthermore, by using nanoscale reinforcements, the loss in breakdown strength can be reduced.

7.5.2.8 *Optical clarity*

PNCs with transparent polymeric matrices are transparent as nano-reinforcement does not cause opacity. Good optical clarity has been obtained in many nanocomposites, particularly at low volume fractions. The reason is attributed to the fact that the size of nanoparticle is lesser than the wavelength of the light. The scattering power for light propagation through a collection of scattering particles can be predicted by Rayleigh scattering:

$$P_{scat} = P_0\, q\, (n' - n)/n^2)\, (V^2/k^4) \qquad (7.5)$$

where P_0 is the incident power, q is the concentration of particles, n′ is the refractive index of the particles, n is the refractive index of the matrix, V is the volume of a single particle and k is the wavelength of light. Therefore, to minimize the scattering, the particles must be as small as possible with an index of refraction as close as possible to that of the matrix material. This equation assumes that the particles are much smaller than the wavelength of light.

7.5.3 *Clay reinforced PNCs*

Clay reinforced nanocomposites have layer structure due to the introduction of polymeric substance into layered inorganic host such as clay [29, 30]. As shown in Figure 7.6, there are two major types of morphological arrangements in clay based polymeric nanocomposites: intercalation and exfoliatation [31]. In the case of intercalation, the organic component is inserted between the layers of the clay such that the inter-layer spacing is expanded, but the layers still bear a well-defined spatial relationship to each other. In this case, polymer chains alternate with the layers of nanoreinforcement in a fixed compositional ratio and have a well-defined number of polymer layers in the intralamellar space. In an exfoliated structure, the layers of the clay are completely separated and the individual layers are distributed throughout the matrix. A third alternative is the dispersion of clay particles (tactoids) within the polymer matrix, where nano clay behaves as a conventional filler and leads to a disordered structure.

Plate-like nanofillers are layered materials typically with a thickness on the order of 1 nm, and with an aspect ratio of at least 25 in other two

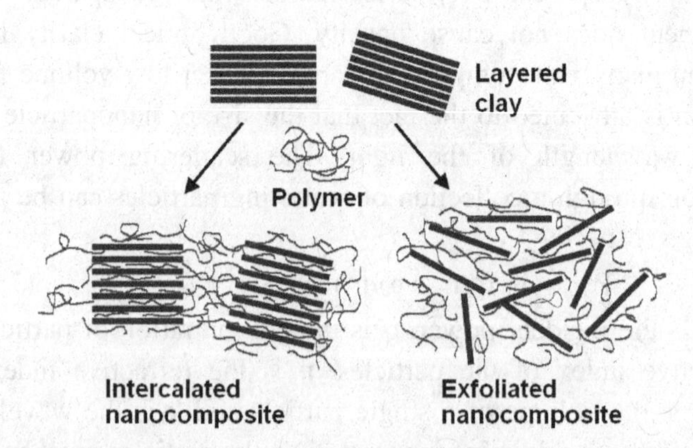

Figure 7.6 Typical morphologies of clay based polymeric nanocomposites [31]. Source: *A Review of Nanocomposites 2000* by J. N. Hay and S. J. Shaw.

dimensions. Three-dimensional (3D) nanofillers are relatively equi-axed particles which are less than 100 nm in their largest dimension. The properties and the fabrication methods of polymeric nanocomposites strongly depend on the geometry, maximum possible dispersion and minimum possible attrition of the nanoreinforcements. Clay reinforced PNCs may be fabricated through *in-situ* polymerization, melt extrusion process and polymer solution mixing process.

7.5.4 *CNT reinforced PNCs*

CNT reinforced PNCs have good strength modulus, thermal stability, solvent resistance, enhancement in glass transition temperature, reduction in thermal shrinkage coupled with good electrical conductivity. In these PNCs, SWNTs are preferred over MWNTs because MWNTs have weak frictional interaction between the layers. The relative sliding between the walls of MWNTs results in variation and mismatch of compressive and tensile loads. Usually, MWNTs experience compressive load on outer wall and tensile force on the inner walls. Nanofiber or nanotube reinforcements have a diameter less than 100 nm and an aspect ratio of at least 100. CNTs may be dispersed in polymer matrix by melt processing, *in-situ* polymerization and solution processing. CNTs reinforced PNCs can be developed into various forms, such as fibers, films and bulk structures.

7.5.5 *Rubber matrix nanocomposites*

Rubber matrix nanocomposites (RMNCs) have major applications as electromagnetic interference (EMI) shields in the field of electronics, and rubber seals and gaskets in high pressure systems. The conducting nanomaterials such as nano silver or CNTs reinforced RMNCs are used as EMI shields. The reinforcement of nano clay improves the barrier properties of the rubber and, hence, they are used as seals and gaskets in high pressure systems. In addition, the reinforcement of nano clay improves the glass transition temperature and loss modulus of the rubber.

7.5.6 *Stealth PNCs*

Polymer based stealth nanocomposites can be used to absorb electromagnetic (EM) radiation and have good mechanical strength, thermal stability and chemical resistance. The metallic surface of a target reflects the EM radiation and this is how RADAR detects the target. A stealth material either absorbs the EM radiation energy or reflects it in chaotic direction, so that RADAR could not receive the back or reflected signal. There are three major properties of a material which qualify it as a stealth material. The first property is the suitable relationship between the dielectric permittivity (ε) and magnetic permeability (μ). The second property is the loss tangent (tan δ). Usually higher loss angle (tan δ) means higher absorbance of the EM radiation energy. The third and the most important property is the reflective loss. Reflective loss is a measure of lost energy when an EM wave passes through the material and reflected back from metallic target through the material again. A good stealth material usually has a high reflection loss over a wide frequency range.

Nanoferrite reinforced nanocomposites are excellent stealth materials. Ferrites are good absorbers for EM radiation. They can be reinforced in low viscous resins such as epoxy resins to make coatings and paints. The mixing, dispersion and wetting of nanoferrite are the key issues. The thickness of a coating is the factor which decides the stealth property for a specific frequency range. Such coatings have a number of applications in marine, aerospace, combat vehicles and so on.

There are high performance thermoplastics, such as PEEK and PES, which can be reinforced by nanoferrite and the resultant composites have good mechanical, chemical strength along with reasonable good stealth properties. Such nanocomposites may be used for structural applications directly. However, the loading of nanoferrite is limited due to high viscosities of polymer melts. Besides, some types of epoxies, such as epoxy resin and polyester resin, reinforced by nanoferrites can be used as stealth paints and coatings in aerospace, civil structures, marine structures, transport vehicles and so on.

7.6 Nano-Bio-Composites

Self-organization and directed assembly of biological macromolecules and inorganic materials play an important role in the creation of the nanostructured and nanocomposite materials in biology. Natural materials unquestionably have exquisite properties that could not be found in synthetic materials. Biological systems can produce these exquisite materials around the room temperature in aqueous environments, whereas most synthetic methods produce materials, which are often inferior to natural biomaterials, at elevated temperature, pressure and with harsh chemicals. Over past decades, materials scientists and chemists have made considerable efforts to create synthetic analogs of biological materials by attempting to mimic biology and, importantly, to learn the design rules of biological systems. Considerable efforts have also been made to create new materials with properties not found in biological systems.

Biological nanocomposites can be entirely inorganic, entirely organic or a mixture of inorganic and organic materials. Though the final material may be entirely one class of material, multiple classes of materials may have been involved in the synthetic process. A good example of a biological nanocomposite in which the organic material does not remain in the final product is the enamel of the mature human tooth, which is 95% by weight hydroxyapatite. During tooth formation, enamel consists of a composite of proteins (primarily amelogenin and enamelin) and hydroxyapatite; however, the proteins are removed as the tooth develops. Figure 7.7 schematically shows an electrophoretic process for hydroxyapatite coating [32].

Typical examples of nanoscale materials in biology include lipid cellular membranes, ion channels, proteins, DNA, actin, spider silk and so forth. In all these structures, the characteristic dimension is on the order of a few nanometers. Although, most of these materials cannot be considered as typical composites, they are composed of discrete nanoscale building blocks, and so most of these biological materials can be considered nanocomposites. In the following, we discuss the nanoparticles synthesized biotanically and biologically.

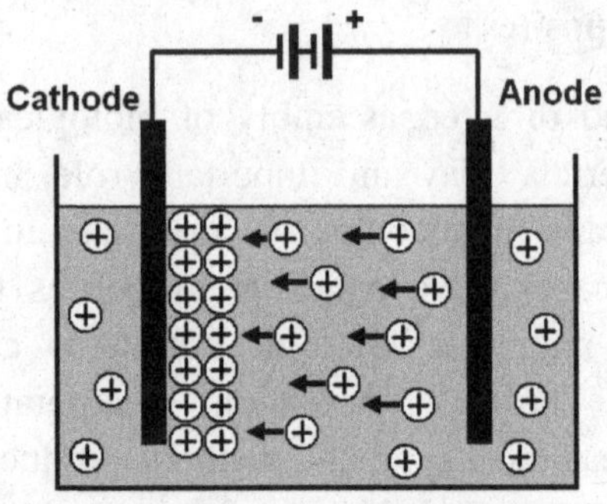

Figure 7.7 Illustration of an electrophoretic process for hydroxyapatite coating [32].

7.6.1 *Botanically synthesized nanoparticles*

For the simplest example of a biological nanocomposite, one needs to look no farther than the grasses. Many species of grasses precipitate SiO_2 nanoparticles within their cellular structures. The sugar cane leaf is an example from which nano SiC can be produced. Perhaps due to some internal structures in their cells, the silica nanoparticles generated within the cells can be found in sheet-like, globular or rod-like morphologies, with characteristic dimensions ranging from a few nanometers to tens of nanometers.

7.6.2 *Biologically synthesized nanostructures*

Biological organisms commonly synthesize nanostructures and nano-composite materials of much greater complexity than nanoparticles. These structures can come in many forms, ranging from needles and plates to complex 3D nanostructures, and can exhibit many unique and interesting mechanical, chemical and optical properties.

Although simplifying all biological nanostructure formation routes into a single mechanism is not possible, several general synthetic tenets tend to hold true. It is generally accepted that the biological structure forms first, and then the inorganic phase begins to form. Although truly cooperative assembly is possible, it is much more common that the organic matrix forms first, regulating the growth of the inorganic material. Usually the organic matrix is plastic and can deform or reshape in response to the growing inorganic material. Besides forming nanoparticles, bacteria can form more complex nanostructures. In general, bacteria are responsible for a vast amount of mineral deposition, and can contribute greatly to the mineral deposits on the bottoms of lakes and other aquatic environments.

7.7 Smart and Intelligent Nanocomposites

Our ability to control the structures and properties of nanocomposites is limited by our understanding of how to manipulate these nanoscale structures. The smart nanocomposites embedded with nanochips and nanosensors, coupled with signal processing and data transfer and display systems, will be able to provide data on stress, strain, heat effect, weatherability, etc. In other words, these structures will be able to forewarn on their health so that preventive and precautionary measures could be taken. Such nanosensor embedded structures can be used in bridges, ship hulls, airframes, SATCOM radomes and so on. Usually the nanosensors are treated with suitable coupling agents, and so they will act as reinforcements due to their nanosizes and the coupling agents. They will remain as an integral part of composites without acting as impurity or source of delamination. Furthermore, if suitable actuation mechanisms are incorporated in these composites and could respond to structure problems with complete repair mechanisms, such composites are usually called intelligent nanocomposites.

Smart and intelligent nanocomposites are the future of materials science and technology. There is a huge market potential for these nanocomposites but the key issue is the cost effectiveness. The advantages offered by nanocomposites outweigh the costs concerns to some extent. Many research and development efforts are being made to develop a variety of nanocomposites to provide the cost effective solution to industries for the development of various products. The efforts in this area are leading to fruitful commercial applications, with significant economic effects.

References:

1. Ajayan, P.M., Schadler, L.S. and Braun, P.V. (2003). *Nanocomposites Science and Technology*, Wiley-VCH Verlag, Weinheim.
2. Iijima, S. (1991). Helical microtubules of graphic carbon, *Nature*, **354**, 56-58.
3. Colbert, D. T. and Smalley, R. E. (2002). Past, present, and future of fullerene nanotubes: Buckytubes, in E. Osawa (ed.), *Perspectives of Fullerene Nanotechnology*, 3-10, Kluwer, Dordrecht, Netherlands.
4. Bacon, R. (1960). Growth, structure, and properties of graphite whiskers, *Journal of Applied Physics*, **31**, 283-290.
5. Ebbesen, T. W. (1997). *Carbon Nanotubes: Preparation and Properties*, CRC Press, Boca Raton, FL.
6. Dresselhaus, M. S., Dresselhaus, G. and Eklund, P. C. (1996). *Science of Fullerenes and Carbon Nanotubes*, Academic Press, New York.
7. Ajayan, P. M. and Ebbesen, T. W. (1997). Nanometre-size tubes of carbon, *Reports on Progress in Physics*, **60**, 1025-1062.
8. Iijima, S. and Ichihashi, T. (1993). Single-shell carbon nanotubes of 1-nm diameter, *Nature*, **363**, 603-605.
9. Jiang, K. Y., Eitan, A., Schadler, L. S., Ajayan, P. M., Siegel, R. W., Grobert, N., Mayne, M., Reyes-Reyes, M., Terrones, H. and Terrones, M. (2003). Selective attachment of gold nanoparticies to nitrogen-doped carbon nanotubes, *Nano Letters*, **3**, 275-277.
10. Bethune, D. S., Kiang, C. H., de Vires, M. S., Gorman, G., Savoy, R., Vazquez, J. and Beyers R. (1993). Cobalt-catalyzed growth of carbon nanotubes with single-atomic-layer walls, *Nature*, **363**, 605-607.
11. Mintmire, J. W., Dunlap, B. I., and White, C. T., Are fullerene tubules metallic, *Physical Review Letters*, **68**, 631-634.
12. Hamada, N., Sawada, S. and Oshiyama, A. (1992). New one-dimensional conductors, *Physical Review Letters*, **68**, 1579-1581.
13. Siegel, R. W., Chang, S. K., Ash, B. J., Stone, J., Ajayan, P. M., Doremus, R. W. and Schadler, L. S. (2001). Mechanical behavior of polymer and ceramic matrix nanocomposites, *Scripta Materialia*, **44**, 2061-2064.
14. Niihara, K., Ishizaki, K. and Isotani, M. (1994). *Materials Processing and Design: Grain-boundary-controlled Properties Of Fine Ceramics II*, American Ceramic Society, Westerville, Ohio.
15. Takacs, L. (1993). Metal-metal oxide systems for nanocomposite formation by reaction milling, *Nanostructured Materials*, **2**, 241-249.
16. Varadan, V. K., Chen, L. and Xie, J. (2008). *Nanomedicine: Design and Applications of Magnetic Nanomaterials, Nanosensors and Nanosystems*, Wiley, Chichester.
17. Bueche, A. M. (1957). Filler reinforcement of silicone rubber, *Journal of Polymer Science*, **25**, 139-149.

18. Kuriakose, B., De, S. K., Bhagawan, S. S., Sivaramkrishnan, R. and Athithan, S. K. (1986). Dynamic mechanical-properties of thermoplastic elastomers from polypropylene natural-rubber blend, *Journal of Applied Polymer Science*, **32**, 5509-5521.
19. Sumita, M., Shizuma, T., Miyasaka, K. and Shikawa, K. (1983). Effect of reducible properties of temperature, rate of strain, and filler content on the tensile yield stress of nylon 6 composites filled with ultrafine particles, *Journal of Macromolecular Science – Physics*, **B22**, 601-618.
20. LeBaron, P. C., Wang, A. and Pinnavaia, T. J. (1999). Polymer-layered silicate nanocomposites: An overview, *Applied Clay Science*, **15**, 11-29.
21. Sumita, M., Tsukumo, Y., Miyasaka, K. and Ishikawa, K. (1983). Tensile yield stress of polypropylene composites filled with ultrafine particles, *Journal of Materials Science*, **18**, 1758-1764.
22. Messersmith, P. B. and Giannelis, E. P. (1994). Synthesis and characterization of layered silicate-epoxy nanocomposites, *Chemistry of Materials*, **6**, 1719-1725.
23. Messersmith, P. B. and Giannelis, E. P. (1995). Synthesis and barrier properties of poly(epsilon-caprolactone)-layered silicate nanocomposites, *Journal of Polymer Science Part A – Polymer Chemistry*, **33**, 1047-1057.
24. Maniar, K. K. (2004). Polymeric nanocomposites: A review, *Polymer – Plastics Technology and Engineering*, **43** (2), 427-443.
25. Shah, V. (1998). *Handbook of Plastics Testing Technology*, 2nd edition, Wiley-Interscience Publication.
26. Gacitua, W. E., Ballerini, A. A. and Zhang, J. (2005). Polymer nanocomposites: Synthetic and natural fillers – a review, *Maderas Ciencia y Tecnología*, **7**(3), 159-178.
27. Ou, Y. C., Yang, F. and Yu, Z. Z. (1998). New conception on the toughness of nylon 6/silica nanocomposite prepared via in situ polymerization, *Journal of Polymer Science, Part B: Polymer Physics*, **36**, 789-795.
28. Drzal, L. T., Rich, M. J., Koenig, M. F. and Lloyd, P. F. (1983). Adhesion of graphite fibers to epoxy matrices. 2. The effect of fiber finish, *Journal of Adhesion*, **16**, 133-152.
29. Yano, K., Usuki, A., Okada, A., Kuraychi, T. and Kamigaito, O. (1993). Synthesis and properties of polyimide clay hybrid, *Journal of Polymer Science Part A – Polymer Chemistry*, **31**, 2493-2498.
30. Kojima, Y., Fukumori, K., Usuki, A., Okada, A. and Kurauchi, T. (1993). Gas permeabilities in rubber clay hybrid, *Journal of Materials Science Letter*, **12**, 889-890.
31. Hay, J. N. and Shaw, S. J. (2000). Clay-based nanocomposites, Abstracted from "*A Review of Nanocomposites 2000*". Retrieved on February 20, 2010 from Azom.com: http://www.azom.com/details.asp?ArticleID=936.
32. Inframat Corporation. http://www.inframat.com/hydro2.htm.

Chapter 8

Organic Electronics

8.1 Introduction

For more than a decade, organic thin film transistors (OTFTs) based on conjugated oligomers, polymers and other molecules have found use in a number of low-cost, large-area electronic applications including flat panel displays, smart cards, radio frequency identification cards (RFIDs), sensors and electronic papers [1-5]. Figure 8.1 shows the world's first flexible, full-color organic light emitting diode (OLED) display built by Sony Electronics Corporation in 2007.

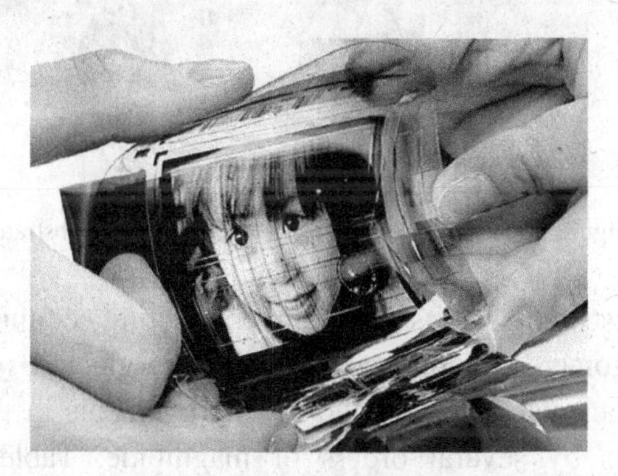

Figure 8.1 The world's first flexible, full-color organic light emitting diode (OLED) display built on organic thin-film transistor (OTFT) technology. Used by permission of Sony Electronics Inc.

The use of organic semiconductors as the active layer in TFTs provide processing advantages over thin film transistors based on inorganic semiconductors such as hydrogenated amorphous silicon (a-Si:H) including reduced processing temperature. Many of these organic thin film materials possess good mechanical properties such as flexibility and toughness, which lead to a low-temperature process using vacuum evaporation, solution casting, ink-jet printing and stamping. This low process temperature allows OTFTs to be integrated on inexpensive substrates of arbitrary size including polymers, enabling roll-to-roll manufacturing, as shown in Figure 8.2. Features that make them particularly attractive are: (a) mouldability, (b) conformability, (c) ease to deposit in the form of thin and thick films, (d) availability of semiconducting and metallic behavior (in selected polymers), (e) the wide choice of their molecular structure and (f) possibility of building charged particles and piezoelectric and pyroelectric effects in the side chain.

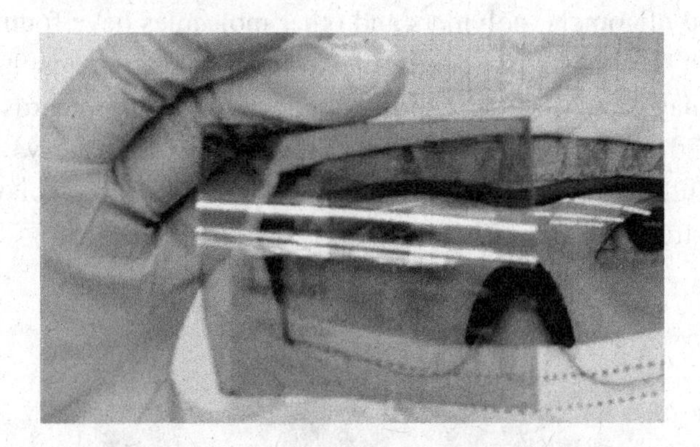

Figure 8.2 Organic electronics on a polymeric substrate.

Recent research efforts on semiconducting conjugated organic thiophene oligomers, thiophene polymers, hexathienyl (α-6T) and the small pentacene molecule have led to improvements in the mobility of these materials by several orders of magnitude. Table 8.1 lists the mobilities of representative organic materials. It may be noted that pentacene has mobility comparable to that of the amorphous silicon. Since these materials are polycrystalline, it may be theoretically impossible to achieve the mobility of the single-crystal silicon. Measurements on single organic crystals of p-type pentacene and an n-type perylene show mobilities of 2.7 cm^2/Vs and 5.5 cm^2/Vs, significantly lower than that of single-crystal silicon. Consequently,

Table 8.1 Field-effect mobility values measured from organic TFT as reported in the literature.

Material	Mobility	Reference	Year
α-ω-hexathiophene	0.03	[7]	1992
α-ω-dihexyl-quinquethiophene	0.1	[8]	2000
α-ω-dihexyl-hexathiophene	0.13	[9]	1996
Dihexyl-anthradithiophene	0.15	[10]	1998
polythienylenevinylene	0.22	[11]	1993
α-ω-dihexyl-quaterthiophene	0.23	[12]	1998
C_{60}	0.3	[13]	1995
α-sexithiophene (α-6T)	0.7	[14]	1998
Pentacene	1.5	[15]	1997
Organic-inorganic hybrid	~1	[2]	2001
Amorphous silicon	~1		
Polysilicon	50-100		
Crystalline silicon	300-900		

OTFTs are not suitable for use in applications requiring very high switching speeds, but they can remain competitive in applications requiring large-area coverage, structural flexibility and low-cost [6].

Pentacene is a well-known p-type material currently under investigation for use as the active layer in TFTs because of its highest field effect mobility of the organic semiconductor reported to date [16]. Since the carrier mobility strongly depends on the π-orbital overlap of neighboring molecules, the ordering of the organic materials controlled by relatively weak van der Waals interactions is the key to the improvement in the mobility. It has been well known that the formation of highly ordered pentacene films resulting in large grains depends on both the growth conditions and the substrate surface such as deposition rate, substrate temperature, substrate morphology and chemical properties of the substrate. Since nucleation of the first monolayer is extremely sensitive to defects of impurities on the substrate, careful treatment of the substrate surface is strictly required to obtain large as well as highly ordered grains. Grain boundaries associated with trap sites act as carrier barriers that dominates charge transport by trapping at the boundaries [17]. This intergrain transport is mostly responsible for the low mobility of the devices. Thus, the investigation of close correlation between the morphology and the structural properties of organic semiconductors should be first required to obtain the most advantageous structures and morphological characteristics for improved performance of organic devices.

Recently, increasing attention has been paid to the use of OTFTs as sensing devices in pursuit of good levels of selectivity, reliability and reproducibility at low cost [18-21]. Such devices employing OTFTs are expected to measure multi-parameters when exposed to chemical species, which may improve the sensor selectivity [21]. Thus, changes in several parameters such as the saturation current, the threshold voltage, the conductivity and the field effect mobility of OTFTs in response to chemical species make them suitable for constructing sensor arrays generally called electronic noses.

Here we discuss the characteristics of pentacene films with respect to their grains and morphology deposited on silicon oxide and silicon at different substrate temperatures for different film thicknesses, which should be first addressed prior to the actual development of various sensors based on pentacene. We also suggest several types of pentacence-based sensors including OTFTs, which could be used for gas, strain, and temperature sensing.

8.2 Morphology of Pentacene Thin Film

Prior to the deposition of pentacene, substrates were prepared from silicon wafers with 3,000 Å of silicon oxide grown by plasma-enhanced chemical vapor deposition (PECVD). Bare silicon was also prepared to compare the surface morphology deposited on both the substrates. Pentacene (Aldrich 97%), without further purification, were thermally evaporated in high vacuum ($\sim 8 \times 10^{-9}$ Torr) at a deposition rate of $0.1 \sim 0.3$ Å/s, keeping the temperature of the source between 150 and 200°C. The substrate was held at temperatures of room temperature, 80°C and 100°C (T_{sub}) with nominal thickness (t_p) up to 5 nm and 50 nm for each deposition temperature. After deposition, a morphological investigation of the pentacene layer was performed by scanning electron microscopy (SEM) as shown in Figures 8.3 and 8.4.

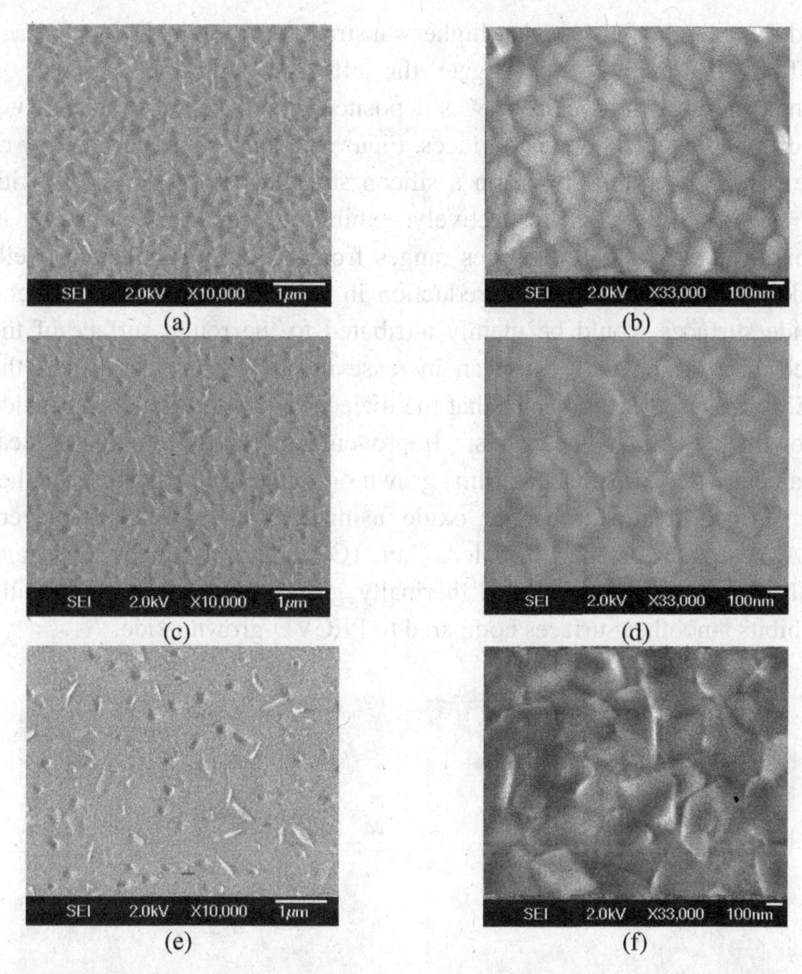

Figure 8.3 SEM images of pentacene film PECVD grown on silicon oxide at a substrate temperature (T_{sub}) and with a nominal film thickness (t_p) of (a) room temperature, 5 nm, (b) room temperature, 50 nm, (c) 80°C, 5 nm, (d) 80°C, 50 nm, (e) 100°C, 5 nm, and (f) 100°C, 50 nm, respectively.

As shown in Figure 8.3, on the oxidized substrates, the grain size increases up to 300 ~ 500 nm at T_{sub} = 100°C from 100 ~ 200 nm at T_{sub} = room temperature with a film thickness of 50 nm, while the roughness of the pentacene film increases with an increase in the substrate temperature. For 5 nm thick films, pin-holes observed in all the films at three different substrate temperatures tend to decrease with temperature, and this may be because that higher substrate temperature favors surface diffuse. In order to investigate the effect of substrate surface on pentacene growth, pentacene was deposited on silicon substrates having hydrophobic and smoother surfaces. Figure 8.4 shows the morphology of the pentacene films grown on a silicon substrate at T_{sub} = 100°C with t_p = 5 nm and 50 nm, respectively, exhibiting a dendritic structure, in which the size of the dendrites ranges from 10~20μm indicating well-ordered film structures. The reduction in crystal size grown on silicon oxide surfaces would be mainly attributed to the rough surface of the dielectric, which may cause an increase in nucleation sites. Thus, the SEM measurement indicates that the dielectrics are too rough to provide growth of larger dentrities. Improvement in the morphological characteristics of pentacene films grown on oxide can be achieved either by surface treatment of the oxide using self-assembled monolayers (SAM) such as octadecyltrichlorosilane (OTS) and hexamethyldisiazane (HMDS), or by employing a thermally grown oxide, which generally exhibits smoother surfaces compared to PECVD-grown oxide.

Figure 8.4 Pentacene on silicon surfaces grown at T_{sub} = 100°C with a film thickness of (a) 5 nm and (b) 50 nm.

8.3 OTFT Fabrication

There are two common device configurations used in organic thin film transistors, as shown in Figure 8.5. The first one is top-contact device

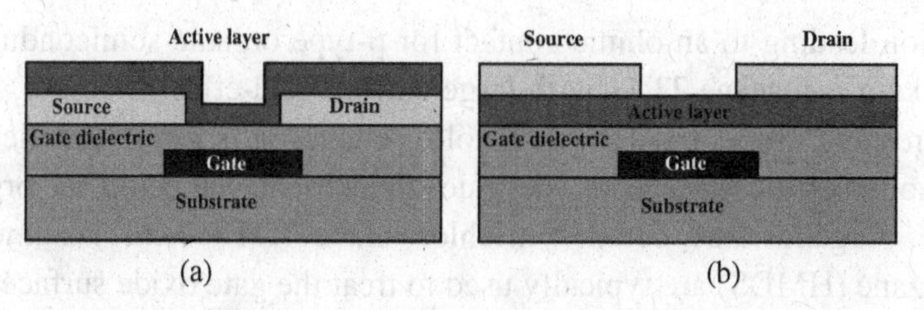

Figure 8.5 (a) Bottom-contact and (b) top-contact devices of organic

with source and drain electrodes evaporated onto the organic semiconducting layer. The other one is bottom-contact device with the organic semiconductor deposited onto prefabricated source and drain electrodes. The performance of pentacene devices with the top-contact configuration is superior to that of bottom-contact devices. In bottom-contact organic TFTs, there are very small crystals of pentacene at the electrode edge, which contain many morphological defects. These defects will create charge-carrier traps with levels lying in the bandgap hence reduce the performance. However, the top-contact devices require shadow masking to pattern the source and drain contacts on top of the pentacene. This is a process that cannot be used in manufacturing. Therefore, the development of a new process enabling the top-contact fabrication should be further explored, which is compatible with standard photolithography.

Figure 8.6 is a process flow for the bottom-contact OTFTs fabricated on a rigid substrate such as glass and silicon. However, they can be demonstrated on cheap, lightweight and flexible polymeric substrates to take full advantage of the potential of organic TFTs and circuits. This require an additional step for mounting flexible substrates such as Kapton or Polyethylene naphthalate (PEN) films on mechanical supports by using a silicone gel adhesive layer.

First, chromium and silicon oxide are deposited and patterned by photolithography to define the gate electrode and gate dielectric, respectively. For the source and drain contact deposited prior to deposition of the active layer, platinum and gold are good candidates because they exhibit a strong resistance to oxidation desirable to avoid the growth of an oxide layer on the metal surface during processing which can act as barrier for charge injection, and possess a high work function leading to an ohmic contact for p-type organic semiconductors. To obtain pentacene TFTs with large carrier field-effect mobility, small subthreshold slope and low threshold voltage, it is often beneficial to functionalize the surface of the gate dielectric layer using an organosilane. Organosilane octadecyltrichlorosilane (OTS) and hexamethyldisilazane (HMDS) are typically used to treat the gate oxide surface prior to pentacene deposition.

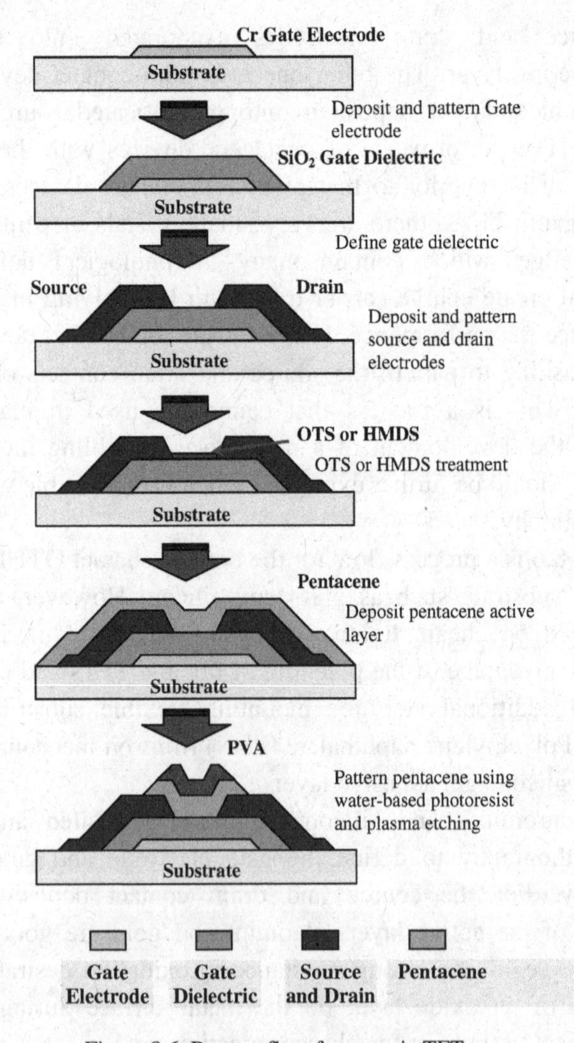

Figure 8.6 Process flow for organic TFTs.

After OTS or HMDS treatment, the pentacene is thermally evaporated to form the active TFT layer in vacuum at a pressure near 10-8 torr with a deposition rate near 1 Å/s. During the pentacene deposition, the substrate is held at 80°C to improve molecular ordering in the pentacene film, which leads to larger carrier mobility and better device characteristics. To pattern the organic active layer, polymer polyvinyl alcohol (PVA) that is a water-soluble polymer is used as photoresist. By using the PVA, the pentacene film can avoid exposure to harmful organic solvents or developers during patterning. The patterned layer of PVA acts as a protective mask while the exposed pentacene is removed by reactive ion etching in oxygen plasma.

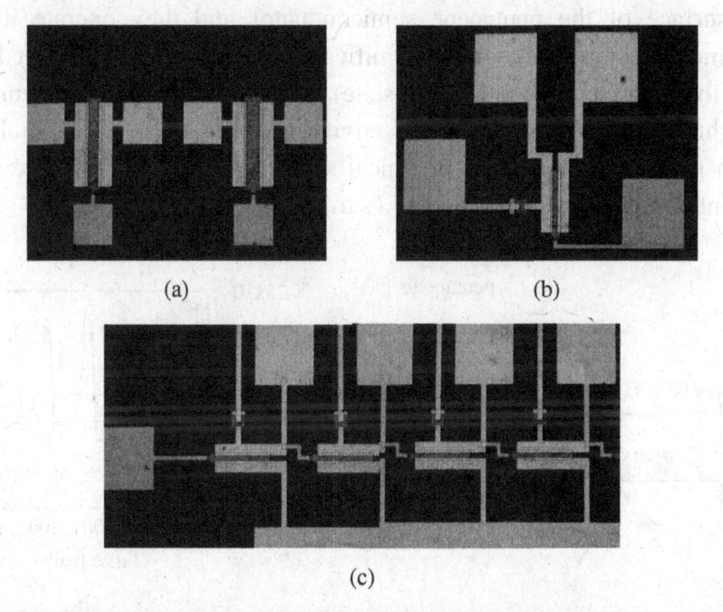

Figure 8.7 Photographs of (a) a TFT with a W/L ratio of 300 μm/40 μm and 300 μm/50 μm, (b) an inverter, and (c) a ring oscillator fabricated on silicon wafers based on pentacene.

Figure 8.7 shows several pentacene circuits fabricated on silicon wafers, which include a pentacene TFT with a W/L ratio of 300 μm/40 μm and 300 μm/50 μm, an inverter and a ring oscillator. The W/L ratio of a TFT is related with the saturation current as the formula shown below:

$$I_{D,sat} = \frac{W \mu C_i}{2L} (V_G - V_T)^2 \quad (8.1)$$

where $I_{D,sat}$ is the saturation current, W is the channel width, L is the channel length, μ is the field effect mobility, C_i is the capacitance of the insulating material per unit area, V_G is the gate voltage and V_T is the threshold. The slope of the plot of the square root of the drain current versus gate voltage relative to the source (V_{GS}) is proportional to the square root of μ. Therefore, from the formula above, the field effect mobility can be extracted.

Polycrstalline pentacene OTFTs exhibit p-type behavior. Thus, when the gate is negatively polarized, an excess of holes will be attracted at the surface of the pentacene semiconductor and they operate in the accumulation mode allowing a significant conductivity to be created in a very thin conducting channel (on state). Change in the saturation current, the threshold voltage, the conductivity and the field effect mobility when pentacene exposes to chemical species, strain and/or temperature variance enables the pentacene TFTs to act as sensing devices.

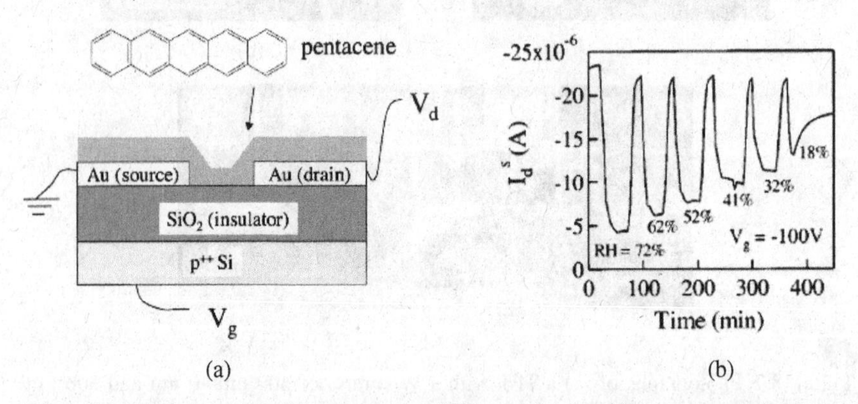

Figure 8.8 (a) Schematic of the pentacene TFT used for humidity sensor, and (b) time dependence of the saturation current measured at a gate voltage of −100V and a drain volatage of −100V when the TFT was exposed to wet N_2 gas with different RH and vacuum, alternatively [20]. Reprinted with permission from Zhu, Z.T., Mason, J.T., Diekmann, R., and Malliaras, G.G., (2002). Humidity sensors based on pentacene thin-film transistors, *Applied Physics Letters*, **81**, 4643-4645. © 2002, American Institute of Physics.

8.4 Typical Organic Sensors

The possibility of using OTFTs for sensing devices has been demonstrated by many research groups. In this case, the conductivity of the organic semiconductor is modulated due to the direct interaction with gas, pressure, humidity and strain. Figure 8.8 shows a schematic of the TFTs used for humidity sensor proposed by Zhu *et al.* [20], and a graph that contains time dependence of the saturation current (I_d^S) when the OTFT was exposed to vacuum and N_2 gas with different levels of relative humidity (RH) alternatively. The saturation current of the OTFT drops dramatically drops by nearly one order of magnitude when exposed to 70% RH.

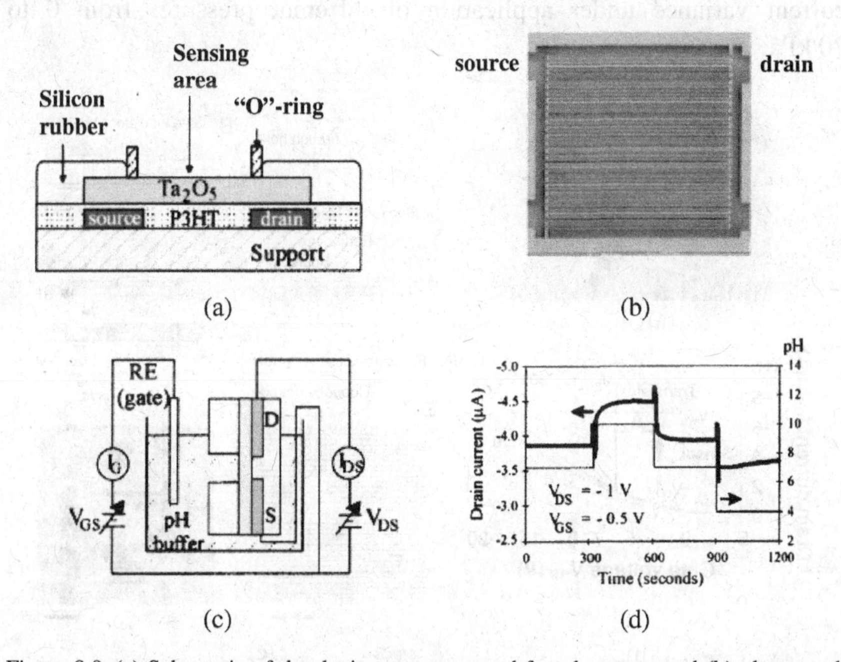

Figure 8.9 (a) Schematic of the device structure used for *ph* sensor, and (b) photograph of the source-drain intedigitated electrodes, (c) scheme of the measurement setup and (d) time dependence drain current corresponding to different *ph* steps [22]. Reprinted with permission from Bartic, C., Campitelli, A., and Borghs, S., (2003). Field-effect detection of chemical species with hybrid organic/inorganic transistors, *Applied Physics Letters*, **82**, 475-477. © 2003, American Institute of Physics.

Bartic *et al.* presented OTFTs fabricated with poly-3-hexylthiphene regio-regular (p3HT) as the semiconductror layer to be used as a ph sensor that can detect both charged and uncharged chemical species in aqueous media as shown in Figure 8.9 [22]. This ph sensor detects ph variations occurring in the bulk of the solution through field-enhanced conductivity of the organic semiconductor using Ta_2O_5 material possessing large numbers of proton-sensitive surface sites. Pressure sensors consisting of OTFTs and pressure-sensitive rubber developed by Someya *et al.* [23] provide the possibility to realize a practical artificial skin thanks to the inherent flexibility and low-cost processing of organic circuits. Figure 8.10 shows the electronic artificial skin device, and the manufacturing process flow of the pressure sensor exhibiting a drain current variance under application of different pressures from 0 to 30 kPa.

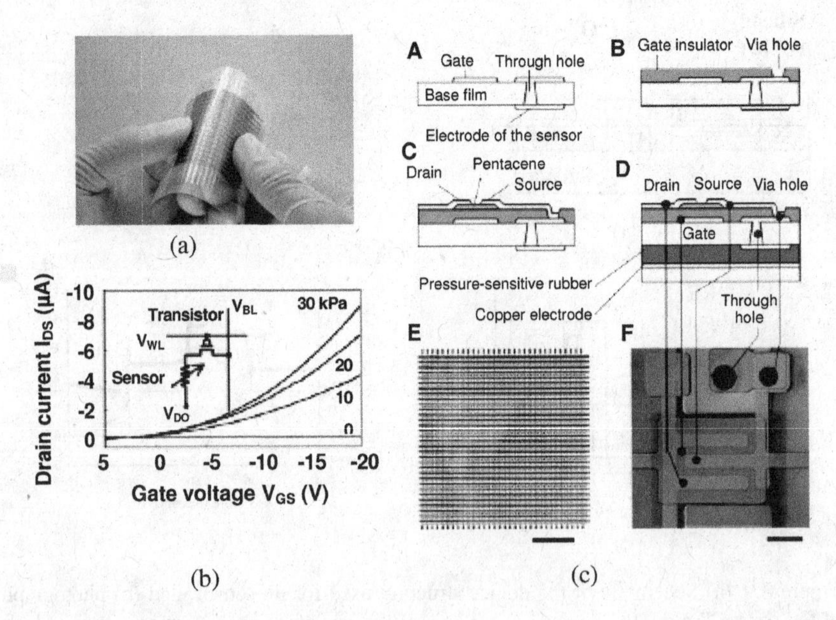

Figure 8.10 (a) Image of an electronic artificial skin, (b) transfer curves (I_{DS} versus V_{GS}) of the pressure sensor under application of various pressures ranging from 0 to 30 kPa and (c) manufacturing process flow [23]. Reprinted with permission from Someya, T., Sekitani, T., Iba, S., Kato, Y., Kawaguchi, H. and Sakurai, T. (2004). A large-area, flexible pressure sensor matrix with organic field-effect transistors for artificial skin applications, *PNAS*, **101**, 9966-9970. © 2004 National Academy of Sciences, U. S. A.

8.5 Strain Sensors

Inorganic semiconductors, for instance, amorphous silicon, have been recently used as a strain sensing element to overcome the limitations of conventional strain sensors based on metallic foil and crystalline silicon for large area sensor applications such as a flexible sensing array for artificial skins of robots, and smart clothing [24-26]. However, the large stiffness mismatch generated between the inorganic semiconductor element and the flexible polymeric substrate may lead to irreversible plastic substrate deformations, and that can be problematic, degrading the sensor performance in terms of reliability and repeatability. The use of organic semiconductors with low Young's modulus (5 GPa) as the sensing element is expected to minimize the induced stress concentration.

Though organic semiconductor strain sensors using pentacene as a sensor active layer has been reported [27], the low conductivity of pentacene, around 10-8 $\Omega^{-1}\cdot cm^{-1}$, would make it difficult to distinguish signal from noise when the sensor operates for practical applications. Recently, there has been reported that carbon nanotubes (CNTs) embedded with organic semiconductor matrix improve the trans-conductance (gm) of organic thin film transistors (OTFTs) [28, 29]. The channel path from source to drain is effectively shortened by the formation of the CNT networks joint by organic semiconductor link, which results in enhancing mobilitiy. Thus, it is expected that employing CNTs-pentacene matrix composite thin film as the active layer of strain sensors improves the signal to noise ratio of sensors.

8.5.1 *Carbon nanotube synthesis*

8.5.1.1 *Apparatus*

Figure 8.11 shows a microwave system for synthesizing CNTs. It consists of a microwave magnetron, a circulator, a four-stub tuner, a waveguide, a cavity, etc. The microwave power can be adjusted from 0 to 3,000 W at a frequency of 2.45 GHz. The function of the circulator is to prevent power being reflected by the load, thus avoiding overheating of the magnetron. The forward and reflected powers are monitored by a power meter which is helpful in determining impedance matching. The four-stub tuner, consisting of four threaded stubs spaced 3/8 of a wavelength apart, is an additional device used to optimize impedance matching. When these stubs are adjusted properly the four-stub tuner becomes a matching network which maximized the power transmitted to the load by matching the source impedance to that of the load. As an important part of the cavity, sliding short is used to adjust the length of the cavity such that it could resonate at 2.45 GHz. High field intensity could be attained when the cavity resonates. A quartz tube, which is used as the reaction chamber, passes through the cavity. Reaction gases are introduced at one end of the quartz tube and exhausted at the other end. The flow rates are controlled by the set of the mass/master flow controller. In this microwave CVD system, SiC is chosen as the substrate because it has moderate loss tangent, and could thus effectively absorb microwave energy.

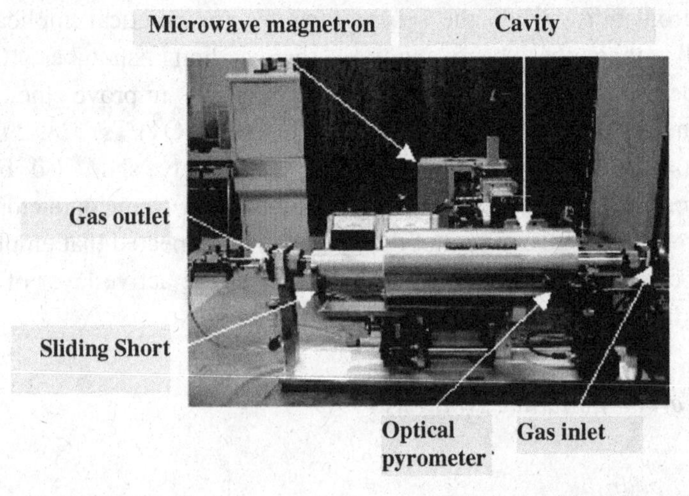

Figure 8.11 Photograph of the actual microwave CVD system for CNT synthesis.

8.5.1.2 *Catalyst preparation*

The catalyst for CNT synthesis is prepared by dissolving iron nitrate in deionized water followed by adding manganese carbonate with vigorous stirring. The obtained semisolid slurry is kept inside a ceramic boat for a thermal anneal process at 400°C overnight. Then the brown-colored solid block after heat treatment is grounded into fine powders by a pastle in a motar.

8.5.1.3 *Microwave CVD synthesis and purification*

A known amount of the catalyst (100 mg) is dispersed on the SiC substrate by a small brush. The substrate is then loaded into the center of the cavity. The Ar gas is purged (100 sccm: standard cubic centimeter per minute) into the cavity and then the microwave is switched on in order to heat the SiC substrate. The temperature is measured and controlled automatically by an optical pyrometer, which is focused on the side of the substrate. Acetylene is introduced into the reactor at a flow rate of 30 sccm at 700°C. The reaction time is set to 30 minutes. After the reaction, the resulting carbon product is removed from the substrate.

A purification process is then conducted to remove the impurities in the as-prepared carbon nanotube materials, which include metal particles, catalyst support and amorphous carbon. It is found that diluted nitric acid is an effective solvent for purification purpose. Thermal gravimetric analysis (TGA) revealed an up to 90% purity of purified carbon nanotubes. Furthermore, to modify the surface properties of the inert carbon nanotube surface, a functionalization process is performed, by which functional groups are attached to the surface of nanotubes. Due to the electrostatic repulsion force among the functional groups, carbon nanotubes can be dispersed relatively well in organic or aqueous solutions.

8.5.1.4 *Electron microscope observation*

To investigate the nanostructure of the prepared carbon nanotubes, transmission electron microscopy (TEM) is performed, typical TEM images are shown Figure 8.12. The diameters of the nanotubes are in the range of up to 20 nm, while their lengths are in micronmeter scale. The high resolution TEM image reveals that the nanotubes have three to five layers of wall, and their outside surfaces are covered by a small

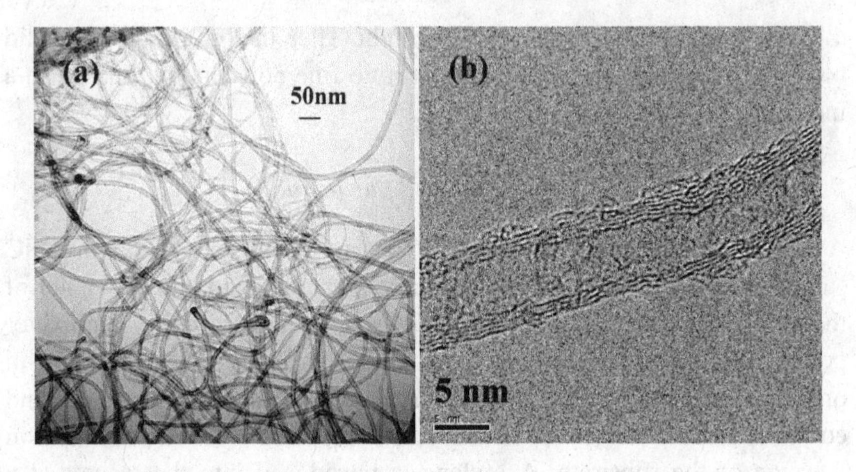

Figure 8.12 TEM images of MWNTs obtained by microwave CVD method (a) low resolution TEM image; and (b) high resolution TEM image.

amount of amorphous carbon. The lattice distance, about 0.34 nm, is consistent with the characteristic of graphitic nanostructures.

8.5.2 *Organic strain sensor fabrication*

Figure 8.13 shows a strain sensor fabricated on Kapton using CNTs-pentacene matrix composites. The fabrication process starts with mounting a Kapton film on a bare silicon carrier with a silicon gel adhesive layer, and a differential pressure vacuum lamination machine is used to facilitate the substrate handling during processes. Solution-processed polyvinlyphenol (PVP) is then spun onto the film in order to make the Kapton surface smoother and relatively hydrophilic. The PVP solution is prepared by mixing PVP powder with poly (melamine-co-formaldehyde) methylated in propylene glycol monomethyl ether acetate (PGMEA). The solution spun on the Kapton is cured at 175°C in a convection oven. A layer of 120 nm thick platinum is then evaporated onto the substrate and defined by lift-off process to form the sensor electrodes. Prior to the deposition of pentacene active layer, the CNTs well dispersed in de-ionized water are spun on, and baked on a hotplate at 100°C for 5 minutes.

Figure 8.13 Strain sensors employing CNTs-pentacene matrix composite as the active sensing layer fabricated on Kapton.

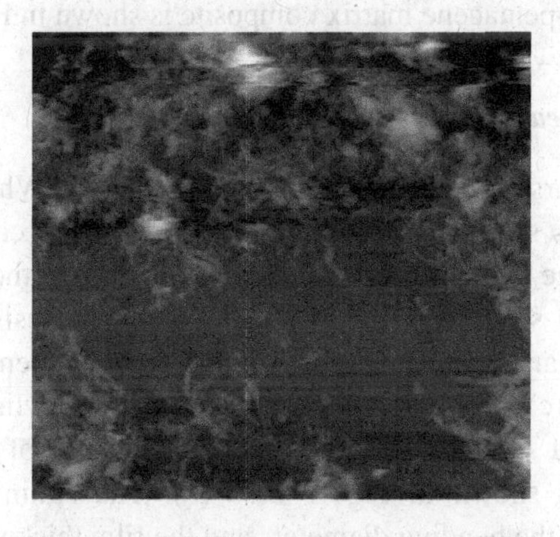

Figure 8.14 AFM image of CNTs spun onto PVP-treated Kapton substrate.

Figure 8.14 shows an atomic force microscopy (AFM) image of the CNTs coated on the PVP-treated Kapton. The substrate temperature is held at 80°C when the pentacene layer is grown at a deposition rate of 0.1-0.3 Å/s up to a 100 nm thickness. The pentacene layer is then patterned by reactive ion etching in oxygen plasma using photosensitized water-soluble polyvinyl alcohol (PVA) as a protective mask. For comparison, two types of strain sensors, with pentacene-CNT composite layer and pentacene single layer respectively, are prepared with an active

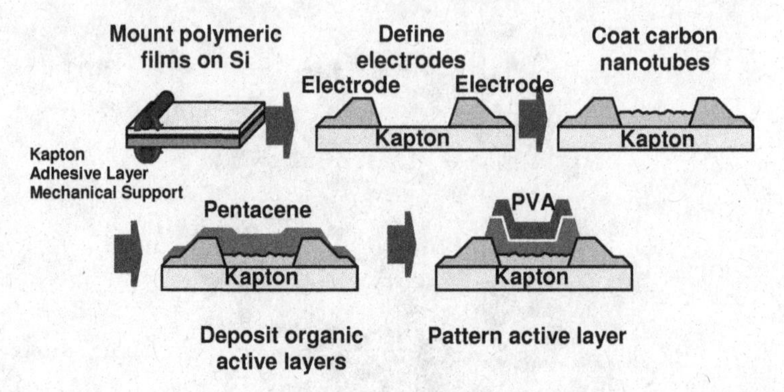

Figure 8.15 Process flow for a strain sensor based on CNTs-pentacene matrix composite thin films.

area of 0.5 mm × 0.5 mm. The whole process flow for the strain sensor based on CNTs-pentacene matrix composite is shown in Figure 8.15.

8.5.3 *Sensor measurement*

Since the sensors developed herein are based on a Wheatstone bridge configuration as shown in Figure 8.16 (a), two metal contacts are used to apply voltage or current, while the other two orthogonal contacts read the output signal varying with the change in resistivity resulting from the mechanical deformation of the sensing element, as shown in Figure 8.16 (c). The Kapton film is bent around cylinders with radii of 50 mm, and 40 mm that correspond to strains of 1, and 1.25%, respectively, as shown in Figure 8.16 (b). The strain values can be calculated from the bending diameters and the film thickness as follows.

$$\varepsilon = \frac{\Delta L}{L} \tag{8.2}$$

$$\Delta L = X - Y = \left(R + \frac{Z}{2}\right)\theta - \left(R - \frac{Z}{2}\right)\theta \tag{8.3}$$

$$L = \left(R + \frac{Z}{2}\right)\theta \tag{8.4}$$

$$\varepsilon \approx \frac{Z}{R} \quad \text{with} \quad R >> Z \tag{8.5}$$

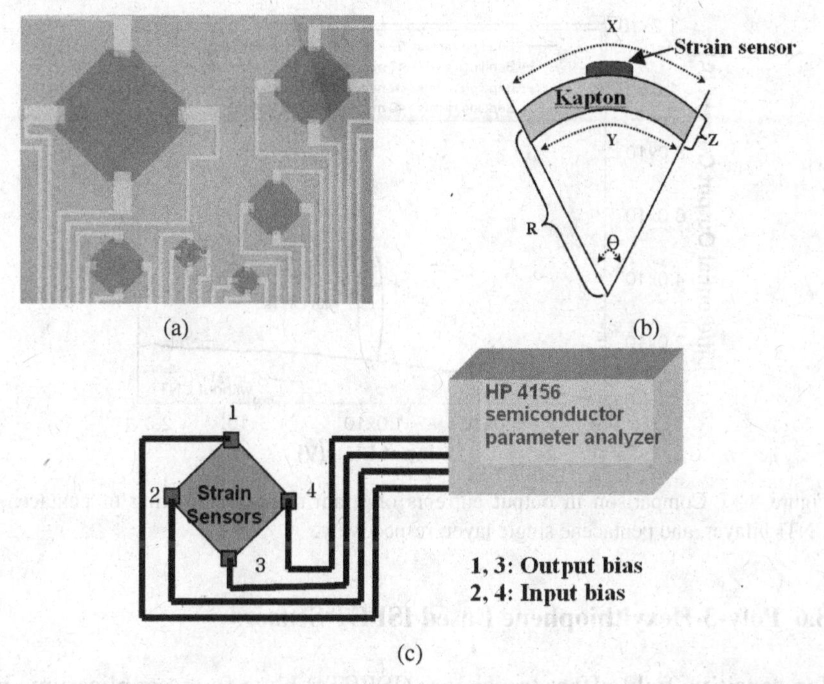

Figure 8.16 (a) Photograph of the pentacene-CNTs strain sensor based on Wheatstone bridge configuration, (b) schematic diagram of strain sensor under bending and (c) measurement setup.

where ε is the strain generated in a sensor, R is the bending radius, X is the outer circular arc, Y is the inner circular arc of a flexible substrate under bending and Z is the substrate thickness, respectively.

Strain is applied at 45° with respect to the direction of the current flow, and an Agilent 4156 semiconductor parameter analyzer is used to apply the bridge bias and to measure the bridge output signals for sensor testing. Figure 8.17 shows the bridge output currents of the pentacene and pentacene-CNTs strain sensors for different bending radii of 40 mm, and 50 mm as a function of input bias. It is interesting that the output signal is substantially enhanced with the addition of CNTs, indicating the improvement in conductivity of the active layer. The output current at 20 V input bias, for instance, is increased by about one order of magnitude with the pentacene-CNTs strain sensor for the bending radius of 50 mm.

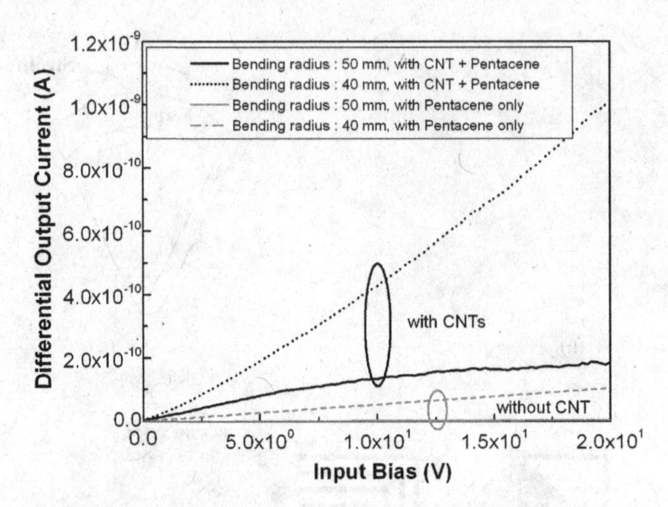

Figure 8.17 Comparison in output currents of strain sensors consisting of pentacene-CNTs bilayer, and pentacene single layer, respectively.

8.6 Poly-3-Hexylthiophene Based ISFET Sensors

Ion-sensitive field effect transistors (ISFETs) have found applications in biomedical engineering as biosensors for *in vitro* and *in vivo* detection purposes. They provide clinicians with cheaper sensors, which have reproducible sensor characteristics that is not possible with the piece-wise assembled sensors used now. Their working is based on the Nernst equation in a thermodynamically well-defined space (cell) [30].

The field effect transistors (FETs) used as biosensors are based on a common principle of converting the immobilized enzyme molecules into charged products. The surface of the semiconductor device being sensitive to these ions gives rise to surface charge which modulates the space charge region at the gate insulator-semiconductor interface. They have the advantage of small size, quick response and the possibility of mass production over the conventional biosensors [31]. Some of the examples are enzyme field effect transistor (EnFET) [32, 33], electrolyte-insulator-semiconductor devices (EIS) [34, 35] and light-addressable potentiometric sensors (LAPS) [36, 37].

These transistors do not have a metal gate electrode, which allows the analyte solution to come in direct contact with the gate dielectric. The potential thus generated induces a charge transport across the source and drain through the semiconductor film. Yan *et al.* attributed the suitability of miniaturizing of the ISFET to the nondependence of their signal to noise ratio on their area [38].

8.6.1 *Biomedical applications*

Usage of ISFET in a flow-injection analysis (FIA) system has been proposed, which eliminates the effects such as signal drift and hysteresis attributed with static measurements with ISFET [39]. The ISFET have found applications as monitoring devices for myocardial ischemia patients. The devices are sensitive to potassium ions (K^+) and hydrogen ions (H^+) also. ISFET has been improvised to show different sensitivity towards the same analyte. A membrane-FET (MEMFET) and a reference-FET (REFET) when used in pair may give a signal independent of the unstable electric potential at the electrode-electrolyte interface [40].

ISFET can also be used as portable urea sensors. The gate dielectric can be functionalized for sensitivity towards NH^{4+} [41, 42] or pH change [43, 44]. Some prototypes also incorporate extended gates. Such prototypes are usually called extended-gate field effect transistors (EGFETs), and they have separate sensitive membranes. Thus detection of urea can be achieved without FET coming in contact with the analyte [45].

Recently, application of the pH-FET is being considered for detection of penicillin-G during production and clinical analysis using Penicillinase. Online sensing application is also tried with this sensor by assessment of the lifespan of the enzyme activity and stability of the enzyme membrane. The performance of the biosensor depends on the type of pH-sensitive material (transducer) used, thickness of the enzyme membrane and on the method of immobilization [31].

8.6.2 *Poly-3-hexylthiophene as a semiconducting polymer*

Solution processed polymers are supposed to exhibit microcrystalline or liquid-crystalline order through self-organization. Regio-regular poly-3-hexylthiophene (p3HT) is a microcrystalline polymer which has been reported to achieve high field-effect mobility of 0.1-0.3 $cm^2V^{-1}s^{-1}$. The microcrystalline structure is an anisotropic laminar arrangement comprising two-dimensional conjugated layers having π-π inter-chain interactions. These interactions are separated by layers of solubilizing, insulating side chains. The result is a fast in-plane charge transport [46], which is sensitive of head-to-tail region-regularity. To enhance or maintain the region-regularity, the substrate surface is pretreated with self-assembling monolayer (SAM), such as hexamethyldisilazene (HMDS) or OTS (octadecyltrichlorosilane) [47]. The mobility is also reported to increase with the increase in molecular weight [48].

Poly-3-hexylthiophene (p3HT) being vulnerable to oxidation and moisture requires thermal treatment which results in de-doping of the film. The thermal treatment (annealing) also affects the On/Off ratio. Hence a thermal cycle has to be chosen considering the trade off between enhancing the mobility of the film and maintaining the On/Off ratio. Literatures suggest that for temperatures less than 60°C, the mobility of the film increases with no significant change in On/Off ratio. At moderate temperatures between 60°C and 120°C, the On/Off ratio increases but mobility degrades with increasing temperature. For higher temperatures both the characteristics are degraded [49]. The selection of a suitable electrode metal is governed by the parameters like electrode-polymer barrier and process considerations like ease of patterning. A barrier height of about 0 eV is reported for p3HT (on HMDS modified surface) with gold as the electrode material [50].

8.6.3 *Tantalum oxide as gate dielectric*

A high dielectric constant for gate dielectric is a viable approach to get high drain currents corresponding to low biases, making the TFT effective even at low stimulus [51]. Tantalum oxide has a very high dielectric constant (between 21 and 24 [52]). The gate dielectric surface is pH sensitive and can act as the gate electrode [53]. The Ta_2O_5 surface exposed to the sample environment gets hydrated by hydrogen ions, hence developing a positive charge.

The quality of the dielectric film is detrimental for the functioning of the TFT. The film should be amorphous and should have a low leakage current (in the order of 10 nanoAmps/cm^2) [54]. The quality of the tantalum oxide film depends on the deposition process. Evaporation, anodizing and sputtering are the processes often used for the deposition of metal oxides. Generally speaking, the sputtering process is preferred over the evaporation process because the former provides a better amorphous layer.

8.6.4 *Enzyme immobilization*

The FET used as a biosensor needs to be functionalized so that it is able to harness the charge induced by the ions (in the analyte) as the gate potential. For this purpose, a layer of immobilized biochemical material (enzyme) is provided on the surface of the gate dielectric. The FET to be used is H^+ sensitive and can be used for qualitative or quantitative sensing by chemical reactions engineered specifically for the substance of interest. The ions (H^+ in this case) are generated by the biochemical reaction between the enzyme and the substance of interest.

Common immobilization techniques use self-assembled monolayers (SAM) [54], poly-vinyl alcohol (PVA) [55] and PVA-SbQ (a derivative of PVA) [56]. In case of SAM, the immobilized enzyme either gets detached over a period of time or losses its activity. The PVA/PVA-SbQ film solves the problem of wear and tear. The resistance of PVA to protein adsorption and cell adhesion provides specificity to the bio-chemical reaction [57]. The Young's modulus of the film is high (in the order of 100 kPa) and increases significantly with the gel concentration for PVA, whereas in case of PVA-SbQ the increase is less [57, 58].

A device can be fabricated in a bottom contact fashion with a tantalum oxide layer sputtered on a p3HT film, as shown in Figure 8.18. A PVA-SbQ hyderogel film with enzyme immobilized can be used as the sensor element which generates the charge for gate potential. The optimum spin coating process and the recipe of the solution is determined based on the surface morphology scans. Special electrode patterns like zigzag inter-digitated electrode are often used to develop miniaturized sensors.

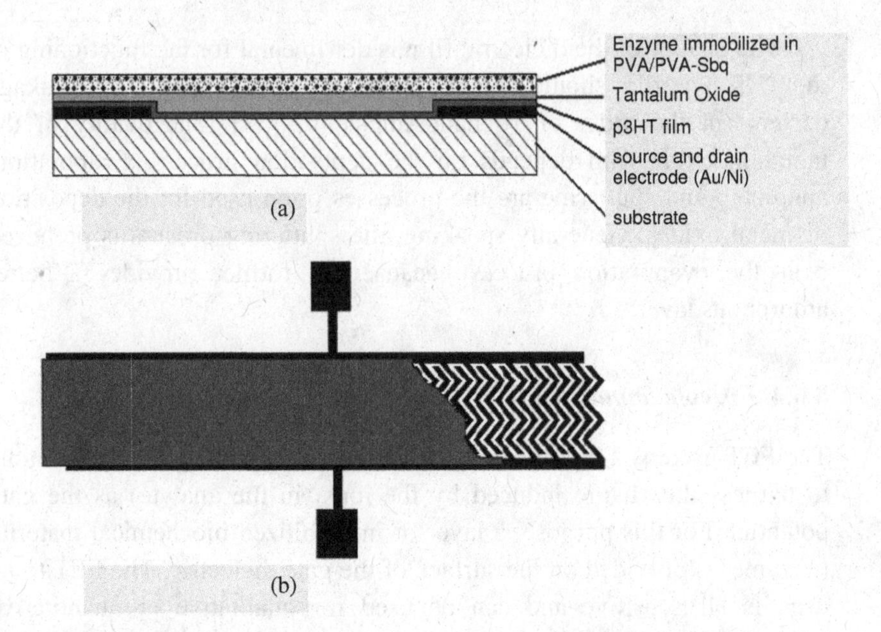

Figure 8.18 (a) The device layout. (b) The zigzag inter-digitated electrode.

8.6.5 *Methodology*

In the fabrication process, three photo mask-layouts are used: a light-field mask for source and drain pattern, a light field mask for p3HT pattern and a dark field mask for UV cross-linkable PVA-Sbq. The widths of the leads and electrodes are not less than 30 μm in order to avoid snapping when the leads wrinkle due to the shrinking of the substrate during the process. The channel lengths is in the range of 10 μm to 50 μm, while the width to length ratio is in the range of 1,000 to 1,500. Apart from the sensors, test patterns with different width to length ratio are fabricated. The contact leads have a pitch of 250 μm which are compatible with the commercially available external contacts. The layers of p3HT and enzyme immobilized films are patterned using separate photo masks giving 5-10 μm margin of error which may be induced by optical lithography. And 0.2, 0.4, 0.8 wt% of p3HT are dissolved in chloroform by allowing it to stand overnight in cool and dark environment. The silicon wafer is coated with HMDS at 150°C. The p3HT solution is spin-coated on the wafer (with and without HMDS) at 1,500 rpm. The p3HT film is then annealed at 120°C for 20 minutes in nitrogen environment, and the thickness of the p3HT film after annealing is several nanometers.

The source and drain are fabricated by deposition of gold using e-beam evaporator and patterning. Usually, to facilitate the adhesion, there is a layer of chromium below the gold.

8.6.6 *Results*

The micrographs of the spin coated film of p3HT show a marked difference in the levels of adhesion and smoothness, as shown in Figure 8.19. The silicon substrate coated with HMDS monolayer provided better adhesion for p3HT molecules. The solution of p3HT in chloroform turns out to have very low viscosity and requires instant spreading over the substrate.

Figure 8.20 shows the atomic force microscopy (AFM) image of the p3HT film that reveals a very smooth surface with micro-aggregate formation.

The zigzag inter-digitated electrodes shown in Figure 8.21 have clear edges. The close proximity of the electrodes warrants special care while doing subsequent processes which involve substantial expansion of the substrate.

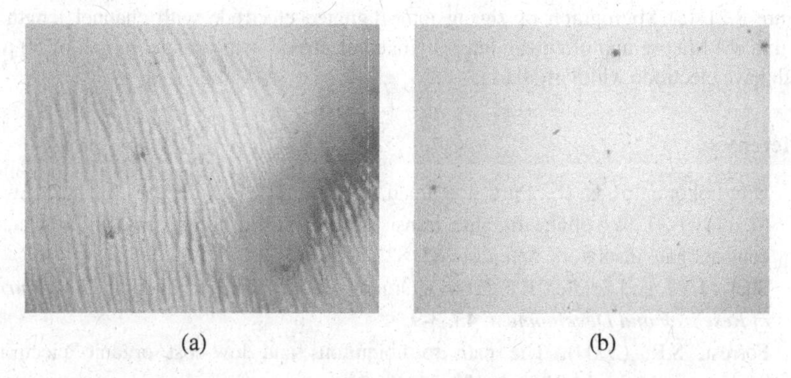

(a) (b)

Figure 8.19 (a) Micrograph of spin coated surface of 0.2 wt% p3HT solution in chloroform on bare silicon wafer at 1,500 rpm. (b) Micrograph of spin coated surface of 0.2 wt% p3HT solution in chloroform on HMDS coated silicon wafer at 1,500 rpm.

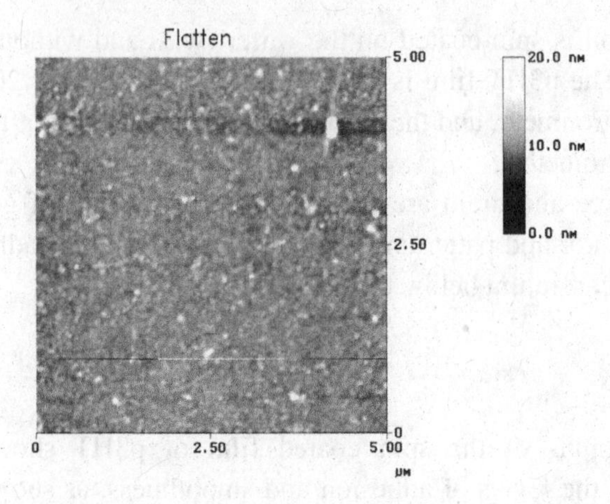

Figure 8.20 AFM image of a spin coated p3HT (0.2 wt%) film.

(a) (b)

Figure 8.21 (a) Micrograph of zigzag inter-digitated electrode with channel length of 20 μm. (b) Micrograph of zigzag inter-digitated electrode with channel length of 50 μm. Both have electrode width of 50 μm.

References:

1. Dimitrakopoulos, C. D., Purushothaman, S., Kymissis, J., Callegari, A. and Shaw, J. M. (1999). Low-voltage organic transistors on plastic comprising high-dielectric constant gate insulators, *Science*, **283**, 822-824.
2. Shaw, J. M. and Seidler, P.F. (2001). Organic electronics: Introduction, *IBM Journal of Research and Development*, **45**, 3-9.
3. Forrest, S.R. (2004). The path to ubiquitous and low-cost organic electronic appliances on plastic, *Nature*, **428**, 911-918.
4. Gelinck, G.H. (2004). Flexible active-matrix displays and shift registers based on solution-processed organic transistors, *Nature Materials*, **3**, 106-110.

5. Baude, P.F., Ender, D.A., Hasse, M.A., Kelly, T.W., Muyres, D.V. and Theiss, S.D. (2003). Pentacene-based radio-frequency identification circuitry, *Applied Physics Letters*, **82**, 3964-3966.
6. Dimitrakopoulos, C.D. and Mascaro, D.J. (2001). Organic thin-film transistors: A review of recent advances, *IBM Journal of Research and Development*, **45**, 11-27.
7. Horowitz, G., Peng, X., Fichou, D. and Garnier, F. (1992). Role of semiconductor/insulator interface in the characteristics of -conjugated-oligomer-based thin-film transistors, *Synthetic Metals*, **51**, 419-424.
8. Katz, H.E., Lovinger, A.J., Johnson, J., Kloc, C., Siergist, T., Li, W., Lin, Y.Y. and Dodabalapur, A. (2000). A soluble and air-stable organic semiconductor with high electron mobility, *Nature*, **404**, 478-481.
9. Dimitrakopoulos, C.D., Furman, B.K., Graham, T., Hegde, S. and Purushothaman, S., (1998). Field-effect transistors comprising molecular beam deposited α-ω-Di-hexyl-hexathienylene and polymeric insulators, *Synthetic Metals*, **92**, 47-52.
10. Laquindanum, J.G., Katz, H.E. and Lovinger, A.J., (1998). Synthesis, morphology, and field-effect mobility of anthradithiophenes, *Journal of the American Chemical Society*, **120**, 664-672.
11. Fuchigami, H., Tsumura, A. and Koezuka, H., (1993). Ploytienylenevinylene thin-film transistor with high carrier mobility, *Applied Physics Letters*, **63**, 1372-1374.
12. Katz, H.E., Lovinger, A.J. and Laquindanum, J.G. (1998). α-ω-Dihexylquarterthiophene: A second thin film single-crystal organic semiconductor, *Chemistry of Materials*, **10**, 457-459.
13. Haddon, R.C., Perel, A.S., Morris, R.C., Palstra, T.T.M., Hebard, A.F. and Fleming, R.M. (1995). C60 thin film transistors, *Applied Physics Letters*, **67**, 121-123.
14. Nelson, S.F., Lin, Y.Y., Gundlach, D.J. and Jackson, T.N. (1998). Temperature-independent transport in high-mobility pentacene transistors, *Applied Physics Letters*, **72**, 1854-1856.
15. Lin, Y.Y., Gundlach, D.J., Nelson, S. and Jackson, T.N., (199&). Stacked pentacene layer organic thin-film transistors with improved characteristics, *IEEE Electron Device Letters*, **18**, 606-608.
16. Klauk, H., Halik, M., Zschieschang, U., Schmid, G. and Radik, W. (2002). High-mobility polymer gate dielectric pentacen thin film transistors, *Journal of Applied Physics*, **92**, 5259-5263.
17. Steudel, S., Vusser, S.D., Jonge, S.D., Janssen, D., Verlaak, S., Genoe, J. and Heremans, P. (2004). Influence of the dielectric roughness on the performance of pentacene transistors, *Applied Physics Letters*, **85**, 4400-4402.
18. Wang, L., Fine, D. and Dodabalpur, A. (2004). Nanoscale chemical sensor based on organic thin-film transistors, *Applied Physics Letters*, **85**, 6386-6388.
19. Crone, B., Dodabalapur, A., Gelperin, A., Torsi, L., Katz, H.E, Lovinger, A.J. and Bao, Z. (2001). Electronic sensing of vapors with organic transistors, *Applied Physics Letters*, **78**, 2229-2231.

20. Zhu, Z.T., Mason, J.T., Diekmann, R. and Malliaras, G.G. (2002). Humidity sensors based o pentacene thin-film transistors, *Applied Physics Letters*, **81**, 4643-4645.
21. Torsi, L., Dodabalapur, A., Sabbatini, L. and Zambonin, P.G. (2000). Multi-parameter gas sensors based on organic thin-film-transistors, *Sensors and Actuators B*, **67**, 312-316.
22. Bartic, C., Campitelli, A. and Borghs, S. (2003). Field-effect detection of chemical species with hybrid organic/inorganic transistors, *Applied Physics Letters*, **82**, 475-477.
23. Someya, T., Sekitani, T., Iba, S., Kato, Y., Kawaguchi, H. and Sakurai, T. (2004). A large-area, flexible pressure sensor matrix with organic field-effect transistors for artificial skin applications, *PNAS*, **101**, 9966-9970.
24. Caputo, D., Cesare, G., Gavesi, M. and Palma, F. (2004). a-Si:H alloy for stress sensor application, *Journal of Non-Crystalline Solids*, **338-340**, 725-728.
25. Zhou, L., Jung, S., Brandon, E. and Jackson, T.N. (2006). Flexible substrate micro-crystalline silicon and gated amorphous silicon strain sensors, *IEEE Transactions on Electron Devices*, **53**, 380-385.
26. Jung, S., Ji, T. and Varadan, V.K. (2006). Point-of-care (POC) temperature and respiration monitoring sensors for smart fabric applications, *Smart Materials and Structures*, **15**, 1872-1876.
27. Bo, X.Z., Tassi, N.G., Lee, C.Y., Strano, M.S., Nuckolls, C. and Blanchet, G.B. (2005). Pentacene-carbon nanotubes: Semiconduting assemblies for thin-film transistor applications, *Applied Physics Letters*, **87**, 2035101-2035103.
28. Kumar, S., Blanchet, G.B., Hybertsen, M.S., Murthy, J.Y. and Alam, M.A. (2006). Performance of carbon nanotube-dispersed thin-film transistors, *Applied Physics Letters*, **89**, 1435011-1435013.
29. Bo, X.Z., Lee, C.Y., Strano, M.S., Goldfinger, M., Nuckolls, C. and Blanchet, G.B. (2005). Carbon nanotube-semiconductor networks for organic electronics: The pickup stick transistor, *Applied Physics Letters*, **86**, 1821021-1821023.
30. Bergveld, P. (2003). ISFET, theory and practice, *IEEE Sensor Conference Toronto*, 1-26.
31. Poghossian, A., Schoning, M. J., Schroth, P., Simonis, A. and Luth, H. (2001). An ISFET-based penicillin sensor with high sensitivity, low detection limit and long lifetime, *Sensors and Actuators B Chemical*, **76**, 519-526.
32. Caras, S. and Janata, J. (1980). Field-effect transistor sensitive to penicillin, *Analytical Chemistry*, **52**, 1935-1937.
33. Osa, T., (1987). Enzyme-based CHEMFET, in: A. P. F. Terner, I. Karube and G. S. Wilson (eds.), *Biosensors Fundamentals and Applications*, Oxford University Press, New York, 481-530.
34. Beyer, M., Menzel, C., Quack, R., Scheper, T., Schugerl, K., Treichel, W., Voigt, H., Ulrich, M. and Ferretti, R. (1994). Development and application of a new enzyme sensor type based on the EIS-capacitance structure for bioprocess control, *Biosensors and Bioelectronics*, **9**, 17-21.

35. Schoning, M. J., Thrust, M., Muller-Veggian, M., Kordos, P. and Luth, H. (1998). A novel silicon-based sensor array with capacitive EIS structure, *Sensors and Actuators B*, **47**, 225-230.
36. Hafeman, D., Parce, J. and McConnell, H. (1998). Light-addressable potentiometric sensor for biochemical systems, *Science*, **240**, 1182-1185.
37. Seki, A., Ikeda, S., Kubo, I. and Karube, I. (1998). Biosensors based on light-addressable potentiometric sensors for urea, Penicillin and glucose, *Analytica Chimica Acta*, **379**, 9-13.
38. Yan, F., Estrela, P., Mo, Y., Migliorato, P. and Maeda, H. (2003). Polycrystalline silicon thin-film transistor ISFETs as disposable biosensors, *204th Meeting of The Electrochemical Society Inc., Abs.* 1210.
39. Rolka, D., Poghossian, A. and Schoning, M. J. (2004). Integration of a capacitive EIS sensor into a FIA system for pH and penicillin determination, *Sensors*, **4**, 84-94.
40. Errachid, A., Ivorra, A., Aguilo, J., Zine, R. and Bausells, J. (2001). New technology for multi-sensor silicon needles for biomedical applications, *Sensors and Actuators B*, **78**, 279-284.
41. Munoz, J., Jimenez, C., Bratov, A., Bartroli, J., Alegret, S. and Dominguez, C. (1997). Photosensitive polyurethanes applied to the development of CHEMFET and ENFET devices for biomedical sensing, *Bioelectronics*, **12**, 577-585.
42. Anne, S., Nicole, J. R., Claude, M. and Serge, C. (1999). A miniaturized urea sensor based on the integration of both ammonium based urea enzyme field effect transistor and a reference field effect transistor in a single, *Talanta*. **50**, 219-226.
43. Jafferzic-Renault, N. and Martelet, C. (1997). Semiconductor based micro-biosensors, *Synthetic Metals*, **90**, 205-210.
44. Kharitonov, A. B., Zayats, M., Linchtenstein, A., Katz, E. and Willner, I. (2000). Enzyme monolayer-functionalized field effect transistors for biosensor applications, *Sensors and Actuators B.*, **70**, 222-231.
45. Chen, J. C., Chou, L. C., Sun, T. P. and Hsiung, S. K. (2003). Portable urea biosensor based on the extended-gate field effect transistor, *Sensors and Actuators B.*, **91**, 180-186.
46. Sirringhaus, H., (2005). Device physics of solution processed organic field-effect transistors, *Advanced Materials*, **17**, 2411-2425.
47. Sirringhaus, H., Tesseler, N. and Friend, R. H. (1998). *Science*, **280**, 1741.
48. Klein, R. J., McGehee, M. D., Kadnikova, E. N., Liu, S. J. and Frechet, J. M. J. (2003). *Advanced Materials*, **15**, 1519.
49. Mattis, B. A., Chang, P. C. and Subramanian, V. (2003). Effect of thermal cycling on performance of poly (3-hexylthiophene) transistors, *Proceedings of Materials Research Society Symposium*, **771**, L10.35.1-L10.35.6.
50. Burgi, L., Richards, T., Friend, R. and Sirringhaus, H. (2003). Close look at charge carrier injection in polymer field-effect transistors, *Journal of Applied Physics*, **94** (9), 6129-6137.

51. Facchetti, A., Yoon, M. H. and Mark, T. J. (2005). Gate dielectrics for organic field-effect transistors: New opportunity for organic electronics, *Advanced Materials*, **17**, 1705-1725.
52. Bartic, C., Jansen, H., Campitelli, A. and Borghs, S. (2002). Ta2O5 as gate dielectric material for low-voltage organic thin-film transistors, *Organic Electronics*, **3**, 65.
53. Gao, C., Zhu, X., Choi, J. W. and Ahn, C. H. A. (2003). Disposable polymer field effect transistor (FET) for pH measurement, *Transductors '03, The 12th International Conference on Solid State Sensors, Actuators and Microsystems*, Boston.
54. Ferretti, S., Paynter, S., Russell, D. A., Sapsford, K. E. and Richardson, D. J. (2000). Self-assembled monolayers: A versatile tool for the formulation of bio-surfaces, *Trends in Analytical Chemistry*, **19** (9), 530-540.
55. Uhlich, T., Ulbricht, M. and Tomaschewski, G. (1996). Immobilization of enzymes in photo chemically cross-linked polyvinyl alcohol, *Enzyme Microbiol Technol.*, **19**, 124-131.
56. Zimmermann, S., Fienbork, D., Stoeber, B., Flouriders, A. W. and Liepmann, D. (2003). A micro needle-based glucose monitor: Fabricated on a wafer-level using in-device enzyme immobilization, *The 12th International Conference on Solid State Sensors, Actuators and Microsystems*, Boston.
57. Schmelden, R. H., Masters, K. S. and West, J. L. (2002). Photo-cross linkable polyvinyl alcohol hydrogels that can be modified with cell adhesion peptides for use in tissue engineering, *Biomaterials*, **23**, 4325-4332.
58. Vogelsang, C., Wijffels, R. H. and Ostgaard, K. (2000). Rheological properties and mechanical stability of new gel-entrapment system applied in bioreactors, *Biotechnology and Bioengineering*, **70** (3), 247-253.

Glossary

Adsorption- Adsorption is a process that occurs when a gas or liquid solute accumulates on the surface of a solid or, more rarely, a liquid (adsorbent), forming a molecular or atomic film (the adsorbate). It is different from absorption, in which a substance diffuses into a liquid or solid to form a solution. The term sorption encompasses both processes, while desorption is the reverse process.

Aerogel- A silicon-based foam composed mostly of air. Often called "frozen smoke" or "blue smoke", aerogels have extremely low thermal conductivity, which gives them extraordinary insulating properties. They are the lowest-density solids known on earth.

Aerosol- A suspension of fine particles (0.01-10 microns) of a solid or liquid in a gas.

Aggregation- A collection of individual units or particles gathered together into a mass or body.

Alkali metals- A group of soft, very reactive elements that includes lithium, sodium and potassium.

Atomic Force Microscopy/Microscope (AFM)- Atomic force microscopy is a technique for analysing the surface of a rigid material all the way down to the level of the atom. AFM uses a mechanical probe to magnify surface features up to 100,000,000 times, and produces 3D images of the surface. The technique is derived from a related technology, called scanning tunneling microscopy (STM). An AFM can work either when the probe is in contact with a surface, causing a repulsive force, or when it is a few nanometers away, where the force is attractive.

Bacteria- Single-celled microorganisms, about one micrometer (one thousand nanometers) across.

Biomimetics- Imitating, copying or learning from nature. The study of the structure and function of biological substances to develop man-made systems that mimic natural ones; imitating, copying or learning from biological systems to create new materials and technologies.

Biopolymer- A polymer found in nature. DNA and RNA are examples of naturally occurring biopolymers. See also **polymer**.

Biosensor- A sensor used to detect a biological substance (for example: bacteria, blood gases or hormones). Biosensors often make use of sensors that are themselves made of biological materials or of materials that are derived from or mimic biological materials.

Biosynthesis- The process by which living organisms produce chemical compounds.

Block copolymers- Self-assembled material composed of long sequences of "blocks" of the same monomer unit, covalently bound to sequences of unlike type.

Bottom up- Building organic and inorganic structures atom-by-atom, or molecule-by-molecule.

Brownian Assembly- Brownian motion in a fluid brings molecules together in various position and orientations. If molecules have suitable complementary surfaces, they can bind, assembling to form a specific structure. Brownian assembly is a less paradoxical name for self-assembly.

Brownian Motion- Motion of a particle in a fluid owing to thermal agitation.

Buckminsterfullerene- A sphere of sixty carbon atoms, also called a buckyball. Named after the architect Buckminster Fuller, who is famous for the geodesic dome that buckyballs resemble.

Buckyball- A popular name for Buckminsterfullerene.

Carbon black- Carbon black is a powdered form of elemental carbon. The primary use of carbon black is in rubber products, mainly tyres and other automotive products, but also in many other rubber products such as hoses, gaskets and coated fabrics. Much smaller amounts of carbon black are used in inks and paints, plastics and in the manufacture of dry-cell batteries.

Carbon nanotubes (CNTs)- CNTs are long and thin cylinders of carbon. These large macromolecules are unique for their size, shape and remarkable physical properties. They can be thought of as a sheet of graphite (a hexagonal lattice of carbon) rolled into a cylinder. The physical properties are still being discovered. Nanotubes have a very broad range of electronic, thermal and structural properties that change depending on the different kinds of nanotube (defined by its diameter, length and chirality, or twist). To make things more interesting, besides having a single cylindrical wall (Single-walled Nanotubes or SWNTs), nanotubes can have multiple walls (MWNTs) — cylinders inside the other cylinders. Sometimes referred to simply as **nanotubes**.

Catalyst- A substance that increases the rate of a chemical reaction by reducing the activation energy, but which is left unchanged by the reaction. A catalyst works by providing a convenient surface for the reaction to occur. The reacting particles gather on the catalyst surface and either collide more frequently with each other or more of the collisions result in a reaction between particles because the catalyst can lower the activation energy for the reaction.

Cell- A small, usually microscopic, membrane-bound structure that is the fundamental unit of all living things. Organisms can be made up of one cell (unicellular; bacteria, for example) or many cells (multi-cellular; human beings, for example, which are made up of an estimated 100,000 billion cells.)

Characterization- Analysis of critical features of an object or concept.

Chemical Vapor Deposition (CVD)- A technique used to deposit thin layers of coatings on a **substrate**. In CVD, chemicals are vaporized and then applied to the substrate using an inert gas such as nitrogen as a carrier. CVD is used in the production of microchips, integrated circuits, sensors and protective coatings.

Chemical vapor transport- A technique similar to CVD used to grow crystal structures.

Chemisorption- The process by which a liquid or gas is chemically bonded to the surface of a solid.

Chirality- The characteristic of a structure (usually a molecule) that makes it impossible to superimpose it on its mirror image.

Chromatography- The physical method of separation in which the components to be separated are distributed between two phases, one of which is stationary while the other moves in a definite direction. Chromatography is widely used for the separation, identification and determination of the chemical components in complex mixtures.

Colloids- Very fine solid particles that will not settle out of a solution or medium. Smoke is an example of a colloid, being solid particles suspended in a gas. Colloids are the intermediate stage between a truly dissolved particle and a suspended solid, which will settle out of solution.

Complementary Metal-Oxide Semiconductor (CMOS)- The semiconductor technology used in the transistors that are manufactured into most of today's computer microchips.

Composite- A material made from two or more components that has properties different from the constituent materials. Composite materials have two phases: matrix (continuous) phase, and dispersed phase (particulates, fibers). For example, steel-reinforced cement is a composite material. The concrete is the matrix phase and the steel rods are the dispersed phase. The composite material is much stronger than either of the phases separately.

Copolymerization- The process of using more than one type of *monomer* in the production of a polymer, resulting in a product with properties different from either monomer.

Crystallography- The process of growing crystals.

Dendrimer- A polymer with multiple branches. Dendrimers are synthetic 3D macromolecular structures that interact with cells, enabling scientists to probe, diagnose, treat or manipulate cells on the nanoscale. From the Greek word dendra, meaning tree.

Dip-pen Nanolithography (DPN)- A method for nanoscale patterning of surfaces by the transfer of a material from the tip of an atomic force microscope onto the surface.

DNA- DeoxyriboNucleic Acid. DNA is a code used within cells to form proteins. A molecule encoding genetic information, found in the cell's nucleus.

DNA structures- DNA frameworks occurring in nature: i.e., double helix, cruciforms, left-handed DNA, multi-stranded structures. Also, microarrays of small dots of DNA on surfaces.

Doping- In electronics, the addition of impurities to a semiconductor to achieve a desired characteristic, often altering its conductivity dramatically. Also known as semiconductor doping.

Drug delivery- The use of physical, chemical and biological components to deliver controlled amounts of a therapeutic agent.

Elastomeric stamp or mould- Key element in soft lithography usually made from polydimethylsiloxane (PDMS), having patterned relief structures on its surface.

Elastomers- Cross-linked high-polymer materials with elastic behavior.

Electrochemical methods- Experimental methods used to study the physical and chemical phenomena associated with electron transfer at the interface of an electrode and a solution. Electrochemical methods are used to obtain analytical or fundamental information regarding electroactive species in solution. Four main types of electrochemical methods include potentiometry, voltammetry, coulometry and conductimetry.

Electrochemical properties- The characteristics of materials that occur when (a) an electric current is passed through a material and produces chemical changes and (b) when a chemical reaction is used to produce an electric current, as in a battery.

Electroluminescence (EL)- The light produced by some materials — mainly semiconductors — when exposed to an electric field. In this process, the electric field excites electrons in the material, which then emit the excess energy in the form of photons. Light-emitting diodes (LEDs) are the most well-known example of EL.

Electron diffraction- A surface science technique used to examine solids by firing a beam of electrons at a sample and observing the electron deflection from the sample's atomic nuclei.

Electron microscopy- The visual examination of very small structures with a device that forms greatly magnified images of objects by using electrons rather than light to create an image. An electron microscope focuses a beam of electrons at an object and detects the actions of electrons as they scatter off the surface to form an image.

Electron transport- The manipulation of individual electrons. Nanolithography techniques allow single electrons to be transported at very low temperatures in specially designed circuits.

Electron tunneling- The passage of electrons through a barrier that, according to the principles of classical mechanics, cannot be breached. An example of electron tunneling is the passage of an electron through a thin insulating barrier between two superconductors. Electron tunneling is a pure quantum mechanical effect that cannot be explained by a classical theory.

Electro-optics- The study of the influence of an electric field on the optical properties of matter — especially in crystalline form — such as transmission, emission and absorption of light. Also known as optoelectronics.

Enzymes- Molecular machines found in nature made of protein, which can catalyze (speed up) chemical reactions.

Epitaxy- The growth of a crystal layer of one mineral on the crystal base of another mineral in such a manner that the crystalline orientation of the layer mimics that of the substrate.

Ferromagnetic materials- Substances, including a number of crystalline materials, that are characterized by a possible permanent magnetization.

Ferromagnetism- A phenomenon by which a material can exhibit spontaneous magnetization. One of the strongest forms of magnetism, ferromagnetism is responsible for most of the magnetic behavior encountered in everyday life and is the basis for all permanent magnets.

Field effect- The local change from the normal value produced by an electric field in the charge-carrier concentration of a semiconductor.

Field emission- The emission of electrons from the surface of a metallic conductor into a vacuum (or into an insulator) under influence of a strong electric field. In field emission, electrons penetrate through the potential surface barrier by virtue of the quantum-mechanical tunnel effect. Also known as cold emission.

Fluorescence- The process in which molecules or matter absorb high energy photons and then emit lower energy photons. The difference in energy causes molecular vibrations.

Fuel cell- An electrical cell that converts the intrinsic chemical free energy of a fuel into direct-current electrical energy in a continuous catalytic process. Fuel cells extract the chemical energy bound in fuel and, in combination with air as an oxidant, transform it into electricity. Researchers are hoping to develop fuel cells that could take the place of combustion engines, thereby reducing the world dependence on fossil fuels.

Fullerene- A molecular form of pure carbon that takes the form of a hollow cage-like structure with pentagonal and hexagonal faces. The most abundant form of fullerenes is C60 (carbon-60), a naturally occurring form of carbon with 60 carbon atoms arranged in a spherical structure that allows each of the molecule's 60 atomic corners to bond with other molecules. Larger fullerenes may contain from 70 to 500 carbon atoms. Named after R. Buckminster Fuller for his writing on geodesic domes; also referred to as **buckyballs**.

Gas-phase reactions- A class of chemical reactions that occur in a single gaseous phase based on the physical state of the substances present. Examples include the combination of common household gas and oxygen to produce a flame.

Green chemistry- The use of chemical products and processes that reduce or eliminate substances hazardous to human health or the environment, creating no waste or generating only benign waste.

Hydrophilic effect- Having an affinity for, attracting, adsorbing or absorbing water. Hydrophilic effect occurs when a liquid comes in contact with another phase — typically a solid substrate, if it attracts the liquid molecules — causing the liquid to attain a relatively large contact area with the substrate.

Hydrophobic effect- Lacking an affinity for, repelling or failing to adsorb or absorb water. Hydrophobic effect occurs when a liquid comes in contact with another phase — typically a solid substrate, if it exerts a repulsive force onto the liquid — causing the liquid to retract from the surface, with relatively little contact area between liquid and substrate.

Infrared (IR) spectroscopy- A technique in which infrared light is passed through matter and some of the light is absorbed by inciting molecular vibration. The difference between the incident and the emitted radiation reveals structural and functional data about the molecule.

Langmuir-Blodgett (LB) films- Ultrathin films (monolayers and isolated molecular layers) created by nanofabrication. An LB-film can consist of a single layer or many, up to a depth of several visible-light wavelengths. The term Langmuir-Blodgett comes from the names of a research scientist and his assistant, Irving Langmuir and Katherine Blodgett, who discovered unique properties of thin films in the early 1900s. Such films exhibit various electrochemical and photochemical properties. This has led some researchers to pursue LB-films as a possible structure for integrated circuits (ICs). Ultimately, it might be possible to construct an LB-film memory chip in which each data bit is represented by a single molecule. Complex switching networks might be fabricated onto multi-layer LB-films chips.

Lattice- In crystallography, a regular periodic arrangement of atoms in three-dimensional space.

LED (Light-emitting Diode)- A semiconductor device that converts electrical energy into electromagnetic radiation. The LED emits light of a particular frequency (hence a particular color) depending on the physical characteristics of the semiconductor used.

LCD (Liquid Crystal Display)- Technology used for displays in notebook and other smaller computers. LCDs allow displays to be much thinner than cathode ray tube technology. LCDs consume much less power because they work on the principle of blocking light rather than emitting it.

Lipids- Lipids are fatty acids and their derivatives, and substances related biosynthetically or functionally to these compounds.

Lithography- The process of imprinting patterns on materials. Derived from Greek, the term lithography means literally "writing on stone." Nanolithography refers to etching, writing or printing at the microscopic level, where the dimensions of characters are on the order of nanometers (units of 10^{-9} meter, or millionths of a millimeter).

Luminescence- Cool light emitted by sources as a result of the movement of electrons from more energetic states to less energetic states. There are many types of luminescence. Chemiluminescence is produced by certain chemical reactions. Electroluminescence is produced by electric discharges, which may appear when silk or fur is stroked or when adhesive surfaces are separated. Triboluminescence is produced by rubbing or crushing crystals.

Macromolecule- A complex large molecule formed from simpler molecules, usually with a diameter ranging from about 100-10,000 angstroms (10^{-5} to 10^{-3} mm).

Macroscale- Larger than nanoscale; often implies a design that humans can directly interact with; too large to be built by a single assembler (one cubic micron of diamond contains 176 billion atoms).

Mass spectrometer- A device used to identify the kinds of molecules present in a given substance: the molecules are ionized and passed through an electromagnetic field. The way in which they are deflected is indicative of their mass and identity.

Matrix- Substance within which something else originates, develops or is contained.

Membrane- In biology, a thin, pliable layer of tissue covering surfaces or separating or connecting regions, structures, or organs of an animal or a plant. In chemistry, a membrane is a thin sheet of natural or synthetic material that can be penetrated, especially by liquids or gases. In environmental applications of nanotechnology, a membrane can be used as a filter.

Mesoporous- Mesoporous materials are porous materials with regularly arranged, uniform mesopores (2-50 nm in diameter). Their large surface areas make them useful as adsorbents or catalysts.

Microelectromechanical Systems (MEMS)- Technology used to integrate various electro-mechanical functions onto integrated circuits. A typical MEMS device combines a sensor and logic to perform a monitoring function. Examples include sensing devices used to control the deployment of airbags in cars and switching devices used in optical telecommunications cables.

Micromachining- The use of standard semiconductor technologies along with special processes to fabricate miniature mechanical devices and components on silicon and other materials.

MOCVD (Metal-Organic Chemical Vapor Deposition)- A technique for growing thin layers of compound semiconductors in which metal-organic compounds are decomposed near the surface of a heated substrate wafer.

Molecular beam epitaxy- Method used to grow layers of materials of atomic-scale thickness on surfaces.

Molecular electronics- Any system with atomically precise electronic devices of nanometer dimensions, especially if made of discrete molecular parts rather than the continuous materials found in today's semiconductor devices.

Molecular imprinting- A process by which functional monomers are allowed to self-assemble around a template molecule and locked into place. The template molecule is then removed, leaving behind a cavity that is complementary in shape and functionality as the template molecule, which will bind molecules identical to the template.

Molecular machine- Any machine with atomically precise parts of nanometer dimensions; can be used to describe molecular devices found in nature.

Molecular manufacturing- Manufacturing using molecular machinery, giving molecule-by-molecule control of products and by-products via positional chemical synthesis.

Molecular nanotechnology- Thorough, inexpensive control of the structure of matter based on molecule-by-molecule control of products and by-products; the products and processes of molecular manufacturing, including molecular machinery.

Molecule- Group of atoms held together by chemical bonds; the typical unit manipulated by nanotechnology.

Monomer- A small molecule that may become chemically bonded to other monomers to form a polymer; from Greek *mono* "one" and *meros* "part".

Nano- A prefix meaning one billionth (1/1,000,000,000).

Nanobiotechnology- Applies the tools and processes of nano/ microfabrication to build devices for studying biosystems.

Nanocomposites- Polymer/inorganic nanocomposites are composed of two or more physically distinct components with one or more average dimensions smaller than 100 nm. From the structural point of view, the role of inorganic filler, usually as particles or fibers, is to provide intrinsic strength and stiffness while the polymer matrix can adhere to and bind the inorganic component so that forces applied to the composite are transmitted evenly to the filler.

Nanocrystalline materials- Solids with small domains of crystallinity within the amorphous phase. Applications include optical electronics and solar cells.

Nanoelectromechanical Systems (NEMS)- A generic term to describe nanoscale electrical/mechanical devices.

Nanoelectronics- Electronics on a nanometer scale, whether made by current techniques or nanotechnology; includes both molecular electronics and nanoscale devices resembling today's semiconductor devices.

Nanofabrication- Design and manufacture of devices with dimensions measured in nanometers.

Nanofibers- Hollow and solid carbon fibers with lengths on the order of a few microns and widths varying from tens of nanometers to around 200 nm.

Nanofluidics- The control of nanoscale amounts of fluids.

Nanolithography- Nanolithography is the art and science of etching, writing or printing of nanoscale patterns. This includes various methods of modifying semiconductor chips at the atomic level for the purpose of fabricating integrated circuits (ICs). Instruments used in nanolithography include the scanning tunneling microscope (STM) and the atomic force microscope (AFM). Both allow surface viewing in fine detail without necessarily modifying it. Either the STM or the AFM can be used to etch, write or print on a surface in single-atom dimensions.

Nanomanipulation- The process of manipulating items at an atomic or molecular scale in order to produce precise structures.

Nanomaterials- Nanoscale particles, films and composites designed and assembled in controlled ways.

Nanoparticles- Particles ranging from 1 to 100 nanometers in diameter. Semiconductor nanoparticles up to 20 nanometers in diameter are often called **quantum dots**, nanocrystals or Q-particles.

Nanoporous materials- Engineered materials with nanoscale holes, used in filters, sensors and diffraction gratings. In DNA sequencing, nanoporous materials have tiny holes that allow DNA to pass through one strand at a time. In biology, complex protein assemblies that span cell membranes allow ionic transport across the otherwise impermeable lipid bilayer.

Nanostructures- Structures made from nanomaterials.

NMR (Nuclear Magnetic Resonance) Spectroscopy- Analytical technique used to determine the structure of molecules. In NMR, the molecule is placed within a strong magnetic field to align the atomic nuclei. An oscillating electromagnetic field is applied, and the radiation absorbed or emitted by the molecule is measured. Not all atoms can be detected using NMR because the nuclei must have nonzero magnetic moments.

Nanomechanical- Being mechanical and very small; for example, a robot that can manipulate single molecules.

Nanomachine- An artificial molecular machine of the sort made by molecular manufacturing.

Nanotechnology- Areas of technology where dimensions and tolerances in the range of 0.1 nm to 100 nm play a critical role. The ability to construct shapes, devices and machines with atomic precision, and to combine them into a wide range of products inexpensively.

Nanotube- A one-dimensional fullerene (a convex cage of atoms with only hexagonal and/or pentagonal faces) with a cylindrical shape.

Nanowires- One-dimensional structures, with unique electrical and optical properties that are used as building blocks in nanoscale devices.

Phase- A part of a sample of matter that is in contact with other parts but is separate from them. Properties within a phase are homogeneous (uniform). For example, oil and vinegar salad dressing contains two phases: an oil-rich liquid, and a vinegar-rich liquid. Shaking the bottle breaks the phases up into tiny droplets, but there are still two distinct phases.

Piezoelectrics- Dielectric crystal that produce a voltage when subjected to mechanical stress or can change shape when subjected to a voltage.

Photolithography- The technique used to produce the silicon chips that make up modern-day computers. The traditional process involves shining light through a mask onto a photosensitive polymer (photoresist) on a silicon surface, then subsequently removing the exposed areas.

Piezoelectricity- The generation of electricity or of electric polarity in dielectric crystals subjected to mechanical stress, or the generation of stress in such crystals subjected to an applied voltage.

Polymer- A macromolecule formed from a long chain of molecules called **monomers**; a high-molecular-weight material composed of repeating sub-units. Polymers may be organic, inorganic or organometallic, and synthetic or natural in origin. See **biopolymer**.

Protein- Large organic molecules involved in all aspects of cell structure and function. The chemical building blocks from which mammalian cells, organs and tissues like muscle are made. Amino acids determine the size, shape and length of protein molecules. They also give protein molecules the odd ability to coil and uncoil like tiny, cellular snakes.

Quantum confinement effect- Atoms caged inside nanocrystals.

Quantum dot- A nanoscale crystalline structure that can transform the color of light. The quantum dot is considered to have greater flexibility than other fluorescent materials, which makes it suited to use in building nanoscale computing applications where light is used to process information. They are made from a variety of different compounds, such as cadmium selenide.

Quantum well- A P-N-P junction in which the "N" layer is ~10 nm (where traditional physics leaves off and quantum effects take over) and an "electron trap" is created.

Quantum wire- Another form of quantum dot, but unlike the single-dimension "dot", a quantum wire is confined only in two dimensions — that is it has "length", and allows the electrons to propagate in a "particle-like" fashion. Constructed typically on a semiconductor base.

Raman spectroscopy- Analysis of the intensity of Raman scattering, in which light is scattered as it passes through a material medium and suffers a change in frequency and a random alteration in phase. The resulting information is useful for determining molecular structure.

Resists- Elements used in performing photolithography experiments. Resists are polymer materials spun onto a substrate. When exposed to UV light, the polymer in the resist cross-links. When treated with a solvent, the cross-linked portion of the resist dissolves, leaving the desired pattern.

Scaffold- Three-dimensional biodegradable polymers engineered for cell growth.

Scanning Electron Microscopy (SEM)- Utilized in medical science and biology and in such diverse fields as materials development, metallic materials, ceramics and semiconductors. SEM involves the manipulation of an electron beam that is scanned across the surface of specially prepared specimens to obtain a greatly enlarged, high-resolution image of the specimen's exposed structure. Specimens are scanned with a very fine probe ("tip") and the strength of interaction between the tip and surface is monitored. The specimen can be observed whole for assessing external structure or freeze-fracture techniques can be used to image internal structures.

Scanning Tunneling Microscope (STM)- A device that obtains images of the atoms on the surfaces of materials — important for understanding the topographical and electrical properties of materials and the behavior of microelectronic devices. The STM is not an optical microscope; instead, it works by detecting electrical forces with a probe that tapers down to a point only a single atom across. The probe in the STM sweeps across the surface of which an image is to be obtained. The electron shells, or clouds, surrounding the atoms on the surface produce irregularities that are detected by the probe and mapped by a computer into an image. Because of the quantum mechanical effect called "tunneling" electrons can hop between the tip and the surface. The resolution of the image is in the order of 1 nm or less.

Self-assembled Monolayers (SAMs)- Monomolecular films that form or self-assemble after immersing a substrate into a solution of an active surfactant.

Self-assembly- At the molecular level, the spontaneous gathering of molecules into well-defined, stable, structures that are held together by intermolecular forces. In chemical solutions, self-assembly (also called Brownian assembly) results from the random motion of molecules and the affinity of their binding sites for one another. Self-assembly also refers to the joining of complementary surfaces in nanomolecular interaction. Developing simple, efficient methods to organize molecules and molecular clusters into precise, pre-determined structures is an important area of nanotechnology exploration.

Semiconductor- A substance, usually a solid chemical element or compound, that can conduct electricity under some conditions but not others, making it a good medium for the control of electrical current.

Simulation- A broad collection of methods used to study and analyze the behavior and performance of actual or theoretical systems. Simulation provides a mechanism for predicting computationally useful functional properties of systems, including thermodynamic, thermo-chemical, spectroscopic, mechanical and transport properties.

Soft lithography- A term for a collection of techniques (nanocontact printing, nanoimprinting, etc.) that are simple in concept and based around nanostructured forms, or moulds.

Sol-gel materials- Gels, glasses and ceramic powders synthesized through the **sol-gel process**; organic-inorganic composite materials.

Sol-gel process- A chemical synthesis technique for preparing gels, glasses, and ceramic powders generally involving the use of metal alkoxides.

Solid-state reactions- Transformations that occur in and between solids and between solids and other phases to produce solids.

Substrate- A wafer that is the basis for subsequent processing operations in the fabrication of semiconductor devices.

Synthesis- Any process or reaction for building up a complex compound by the union of simpler compounds or elements.

Template- In cell and molecular biology, the macromolecular model for the synthesis of another macromolecule.

Template synthesis- The engineered design and creation of materials of controlled size, shape and surface chemistry.

Transmission Electron Microscopy (TEM)- The use of electron high-energy beams to achieve magnification close to atomic observation.

Thin films- Films having thickness less than 10 micrometer.

Top down- Refers to making nanoscale structures by machining and etching techniques.

UV/VIS (Ultraviolet-Visible) Spectroscopy/Spectrophotometry- Method to determine concentrations of an absorbing species in solution. This technique uses light in the visible and adjacent near ultraviolet (UV) and near infrared (NIR) ranges to achieve this quantitative analysis.

Virus- A parasite (consisting primarily of genetic material) that invades cells and takes over their molecular machinery in order to copy itself.

Wetting- In electronics, coating a contact surface with an adherent film of mercury. In metallurgy, wetting refers to spreading liquid filler metal or flux on a solid base metal. Wetting occurs if a liquid is in contact with another phase, typically a solid substrate, with the substrate exerting an attractive force on the liquid molecules.

X-ray analysis- The use of X-ray radiation to detect heavy elements in the presence of lighter ones, to give critical-edge absorption to identify elemental composition and to identify crystal structures by diffraction patterns.

X-ray Diffraction (XRD)- The scattering of X-rays from a crystal, resulting in an interference pattern used to determine the structure of the crystal.

Zeta potential- Surface charge on nanomaterial.

Index